探索
机器人世界

ROS2 编程入门

A Concise Introduction to Robot
Programming with ROS2

[西] 弗朗西斯科·马丁·里科 著
（Francisco Martín Rico）

黎声 邵帅 万学凡 高幸 朱伟章 译

机械工业出版社
CHINA MACHINE PRESS

A Concise Introduction to Robot Programming with ROS2 by Francisco Martín Rico (9781032264653).

Copyright © 2023 Francisco Martín Rico.

Authorized translation from the English language edition published by CRC Press, part of Taylor & Francis Group LLC. All rights reserved.

China Machine Press is authorized to publish and distribute exclusively the Chinese (Simplified Characters) language edition. This edition is authorized for sale in the Chinese mainland (excluding Hong Kong SAR, Macao SAR and Taiwan). No part of this publication may be reproduced or distributed in any form or by any means, or stored in a database or retrieval system, without the prior written permission of the publisher.

Copies of this book sold without a Taylor & Francis sticker on the cover are unauthorized and illegal.

本书原版由 Taylor & Francis 出版集团旗下 CRC 出版公司出版，并经授权翻译出版。版权所有，侵权必究。

本书中文简体字翻译版授权由机械工业出版社独家出版并仅限在中国大陆地区（不包括香港、澳门特别行政区及台湾地区）销售。未经出版者书面许可，不得以任何方式复制或抄袭本书的任何内容。

本书封底贴有 Taylor & Francis 公司防伪标签，无标签者不得销售。

北京市版权局著作权合同登记　图字：01-2023-2144 号。

图书在版编目（CIP）数据

探索机器人世界：ROS2 编程入门 /（西）弗朗西斯科·马丁·里科著；黎声等译．－－北京：机械工业出版社，2025.6．－－ISBN 978-7-111-78582-8

I. TP242

中国国家版本馆 CIP 数据核字第 2025TQ4213 号

机械工业出版社（北京市百万庄大街 22 号　邮政编码 100037）
策划编辑：王　颖　　　　　　　　责任编辑：王　颖　刘松林
责任校对：赵玉鑫　王小童　景　飞　责任印制：任维东
河北鹏盛贤印刷有限公司印刷
2025 年 8 月第 1 版第 1 次印刷
165mm×225mm・15.75 印张・6 插页・305 千字
标准书号：ISBN 978-7-111-78582-8
定价：79.00 元

电话服务　　　　　　　　　　网络服务
客服电话：010-88361066　　　机　工　官　网：www.cmpbook.com
　　　　　010-88379833　　　机　工　官　博：weibo.com/cmp1952
　　　　　010-68326294　　　金　书　网：www.golden-book.com
封底无防伪标均为盗版　　　　机工教育服务网：www.cmpedu.com

THE TRANSLATOR'S WORDS
译者序

随着科技的飞速发展,机器人技术在现代工业和日常生活中发挥着越来越重要的作用。在这个背景下,特斯拉推出的 Optimus 机器人系列,尤其是第二代 Optimus 机器人(Optimus Gen 2),展示了人形机器人在技术上的突破和在实际应用上的潜力。Optimus Gen 2 在行走速度、平衡性和操作精度上有了显著提升,它在制造工厂中应用越来越广泛,这为机器人在各行业的广泛使用铺平了道路。

在这一背景下,本书显得尤为重要,机器人操作系统(ROS)已经成为机器人编程领域的标准。作为一个开源框架,ROS 极大地简化了机器人开发,使得研究人员和工程师能够专注于创新和应用的实现。ROS2 是 ROS 的升级版本,带来了更加稳定和高效的性能,满足了现代机器人领域的技术需求。

本书内容深入浅出,为读者提供了系统学习 ROS2 的方法。本书由 Francisco Martín Rico 博士撰写,他在机器人领域拥有丰富的经验和深厚的知识背景。他介绍了 ROS2 的核心概念和实际应用,内容涵盖了从基础的机器人编程技巧到复杂的机器人行为建模。本书提供了详尽的示例代码和项目设计,这极大地提升了本书的实用性与可操作性。通过这些内容,读者不仅可以学习 ROS2 的基本概念和编程技巧,还有机会在实践中不断深化理解并应用这些知识。

本书的每一个实例和代码片段都是精心设计的,旨在帮助读者在实践中学习。在翻译过程中,我们在保持原书风格和准确性的同时,结合了自身的实践经验,参阅了大量的相关资料,对一些技术细节进行了注释,以更好地帮助读者理解内容。

希望本书能为读者提供一份既准确又实用的学习资源。

凯捷中国数字化团队的王星、贺敏思、李晨源、黄训奇、付高杨、姜雪杰、龚雪云共同参与了本书的审校,他们的帮助使本书更具可读性,在此对他们表示感谢。

机器人技术是前沿科技领域的重要组成部分,掌握先进的机器人编程技术至关重要。愿本书能够激发更多人对机器人技术的兴趣和热情,共同推动机器人技术的发展。

FOREWORD
推荐序一

本书不仅是ROS2编程的入门指南，更是一把打开未来智能生活的钥匙。它深入浅出地介绍了ROS2——机器人操作系统的新版本，为读者提供了从基础概念到实际应用的全面知识。本书不仅适合工程领域的专业人士学习，也适合那些希望将机器人技术融入商业战略的领导者阅读。

在清华大学EMBA的学习过程中，我们积极探索跨界融合的潜力，主动思考如何将前沿技术与商业实践有机结合。本书也体现了这样的理念。本书不仅详细介绍了ROS社区、计算图和工作空间，还通过实际的编程项目，让读者理解如何将理论应用于实践。ROS2不仅具有灵活性和扩展性，还倡导模块化和标准化。模块化和标准化能够实现机器人技术的广泛应用和快速迭代。本书为我们提供了实现这一愿景的工具和方法。

本书的作者Francisco Martín Rico博士以丰富的经验和深刻的见解，揭示了ROS2编程的精髓。从基础的机器人编程到复杂的机器人行为建模，每一章的内容都是精心设计的，旨在帮助读者构建坚实的机器人编程基础。我强烈推荐本书，它适用于所有对未来智能技术充满好奇和热情的人。我相信这些知识不仅在技术层面至关重要，也是理解和把握未来商业趋势的关键。我诚挚地邀请大家一起探索书中的奥义。

龙军，感进机器人（深圳）有限公司创始人

FOREWORD
推荐序二

科技的发展日新月异，机器人技术已经成为推动社会进步的重要力量。

本书是 ROS2 机器人编程的入门书籍，由 Francisco Martín Rico 博士撰写，他不仅是机器人领域的专家，也是 ROS 社区中享有盛誉的成员。本书不仅详细介绍了 ROS2 的基本概念和工具，还通过实际的编程项目，使读者能够快速掌握 ROS2 的核心技能。我始终认为，技术的力量在于其能够解决实际问题，提高效率，ROS2 正是这样一项技术。

在清华大学 EMBA 的课程中，我们经常讨论如何将前沿技术融入商业实践。以擦窗机器人为例，这种机器人能够自动清洁高层建筑的窗户，不仅提高了清洁效率，还大大降低了人工清洁的风险。本书的实践项目，如行为树和 Nav2 导航等，都是开发此类机器人的关键技术。本书不仅适合机器人相关专业的学生和工程师学习，也适合像我这样的商业管理学生学习。本书语言简洁，实例丰富，即使是编程新人也能轻松上手。

我相信，通过阅读这本书，你将能够更深入地理解 ROS2，并将其应用于各类机器人项目，从而能够解决实际问题，创造更大的商业价值。我强烈推荐本书给所有对机器人技术感兴趣的读者，让我们一起探索机器人科技世界的无限可能。

牛立群，山西嘉世达机器人技术有限公司董事长

FOREWORD
推荐序三

在数字化时代，自动化和智能化已成为推动行业进步的关键因素。在仓储物流领域，仓储机器人技术，以高效、智能、柔性的特点，引领着行业的变革。本书不仅深入浅出地介绍了 ROS2 的编程知识，还提供了一个与机器人创新技术相结合的实践平台，这让我们能够更好地理解和应用这些前沿技术。

在仓储自动化领域，机器人的灵活性和智能化水平是提高效率的关键。借助 ROS2，我们可以为机器人开发复杂的行为树和导航算法，使它们能够在复杂的仓储环境中自如地进行货物搬运和路径规划。本书详细介绍了如何使用 ROS2 实现这些功能，对于希望提升仓储自动化水平的企业来说，这是极具价值的。

本书不仅适合那些从事机器人编程的工程师阅读，也适合企业决策者和技术管理者阅读。它不需要读者具备 ROS2 的经验，而是从零开始，逐步引导读者进入 ROS2 的世界。本书提供的代码为读者提供了实践的机会，读者可以通过实际操作来加深对相关概念的理解。这种理论与实践相结合的方式会提高学习效率，帮助读者更快地掌握 ROS2 编程。对于企业来说，这种方式尤其重要，因为它能够帮助企业快速培养能够实际应用 ROS2 编程的人才。

感谢凯捷中国数字化团队的技术专家为我们带来了这样一本好书，我们正在一起探索机器人技术的无限可能，为企业的数字化转型和智能化升级提供强有力的技术支持。祝开卷有益。

郑睿群，海柔创新科技有限公司研发团队负责人

FOREWORD
推荐序四

从工业制造的精密操作，到医疗领域的精细治疗，再到物流配送的高效运转，机器人以卓越的性能和独特的作用，正在重新塑造着我们的工作模式和生活场景。机器人技术也正以破竹之势成为各行各业变革的先锋力量。ROS2 这一技术为机器人技术开启了崭新的篇章，而本书正是我们深入探索 ROS2 世界的指南。

在全球范围内，ROS2 的发展前景无比广阔。随着人工智能、大数据、云计算等前沿技术的突飞猛进，ROS2 的智能化潜力得到了前所未有的扩展。作为机器人操作系统的中坚力量，ROS2 为机器人的智能化发展奠定了坚实的基础。众多顶尖的机器人公司和研究机构都在积极采用 ROS2 进行机器人的开发和研究，不断使 ROS2 技术更加完善。例如，在工业机器人领域，ROS2 的应用可以极大地提升机器人的运动控制、感知能力和决策效率，从而提高生产效率和产品质量；在服务机器人领域，ROS2 的应用可以实现机器人的自主导航、人机交互和智能服务，从而极大地提升服务能力和用户体验；在医疗机器人领域，ROS2 的应用可以精确控制手术机器人的运动和操作，从而显著提高手术的精确度和安全性。

掌握 ROS2 编程知识，将有助于推动机器人技术的创新与应用。感谢凯捷中国数字化团队的工程师，我衷心期盼本书能吸引更多有志之士投身于机器人技术的研发与探索，共同推动机器人技术向更高水平迈进。

卢林，追觅科技首席信息官

PREFACE
前　言

本书旨在带领读者踏上 ROS2 机器人编程之旅，展示了主要应用 ROS2 概念的若干个项目。读者无须提前掌握 ROS/ROS2 知识。对于 ROS1 程序员而言，他们对本书介绍的许多概念会非常熟悉。ROS2 相较其前代版本的变化会更有趣。

本书的首选语言是 C++，但我们的第一个例子也提供了 Python 语言实现。我们可以使用 Python 开发复杂和强大的项目，但我更喜欢使用编译语言而不是解释语言。同样，用 C++ 来解释概念和用 Python 来解释是一样的。另外，本书使用的是 Linux（具体来说是 Ubuntu GNU/Linux 22.04 LTS），而不是 Windows 或 Mac，因为前者是 ROS2 官方参考平台且在机器人领域占主导地位。

本书内容涉及了许多 C++ 特性，包括智能指针（shared_ptr 和 unique_ptr）、容器（vector、list、map）、泛型编程等，还涵盖了 C++ 17。我将尽力从语言角度解释复杂的代码部分，读者也可查阅一些参考资料⊖⊖。我还期望读者了解 CMake、Git、gdb 和其他在 Linux 环境中使用的常见工具。如果你还不了解这些工具，阅读本书就是学习它们的好机会。

建议读者按照章节顺序阅读本书，如果跳过了某些章节，对于 ROS2 的初学者来说，则会难以理解后续概念。在一些地方，我会插入一个如下所示的文本框：

> **深入了解：某一主题**
>
> 解释说明部分。

这个文本框表示在第一次阅读时可以跳过，并在以后深入理解某些概念时返回查看。本书会使用 shell 命令。ROS2 主要在 shell 中使用，掌握 shell 这样的命令对

⊖ https://en.cppreference.com/w。

⊖ https://www.cplusplus.com。

用户来说非常重要。我将使用这些文本框来显示终端中的命令：

```
$ ls /usr/lib
```

本书并非一本传统的 ROS2 教程。官方网站上的教程已经很好了！事实上，许多概念（服务和动作）在这些教程中学习效果最佳。本书的目标是通过将这些概念应用于机器人执行任务的示例中来帮助读者更好地理解和掌握 ROS2。此外，读者通过本书还能够学习机器人学中的一般概念以及它们在 ROS2 中的具体应用。

因此，在本书中，我们将深入分析大量的代码。为此，我已经准备了一个代码仓库，其中包含了本书将使用到的所有代码⊖。

在每章的结尾，我会提供"建议的练习"来帮助加强你对这个主题的理解。如果你成功解决了它，那么可以将其上传到本书的官方代码库，以一个独立的分支形式附上说明和你的署名。你可以通过向官方书籍代码仓库发起拉取请求（Pull Request，PR）来完成此操作。如果我有时间（希望如此），我会很乐意对其进行评审并与你讨论。

为了让本书更完整，我在附录中添加了这些软件的全部源代码。当涉及说明一个软件包的结构时，将用到此框：

```
Package my_package

my_package/
├── CMakeLists.txt
└── src
    └── hello_ros.cpp
```

为了展示源代码，将使用下面的框：

```
src/hello_ros.cpp

#include <iostream>
int main(int argc, char * argv[]) {
        std::cout << "hello ROS2" << std::endl;
        return 0;
    }
```

此外，如果是为了展示代码片段，那么我将使用下面这种不带编号的方框：

```
std::cout << "hello ROS2" << std::endl;
```

我希望你喜爱本书，让我们开始 ROS2 编程机器人的旅程吧！

⊖ https://github.com/fmrico/book_ros2。

ACKNOWLEDGEMENTS
致　谢

我要感谢那些使本书成为可能的人们。

首先，我要感谢 Steve Macenski 和 Michele Collendise，他们对我寄予厚望。

其次，我要感谢 Fran 和 Vicente，他们分别向我介绍了 ROS 和机器人学，并且一直是我在科学世界和大学中的最好旅伴。

我想特别提到 ROS 社区的成员们，他们投入了大量的时间和精力来改进本书，并发送了评论和修正意见。Andrej Orsula、Sam Pfeifer 和 Zahi Kakish 给我发送了详尽、高质量的评论，这极大地提高了本书的质量。Varun Vivek Vennavalli、Brian Hope 和 Jorge Turiel 也提供了修正意见。

最后，我要感谢我的父母，他们支持我完成学业，这对他们来说意味着巨大的努力。我还要感谢 Marta 在本书写作过程中给予的帮助。

Francisco Martín Rico

目 录

译者序
推荐序一
推荐序二
推荐序三
推荐序四
前言
致谢

第1章 引言 ················· 1
1.1 ROS2 概述 ················ 2
   1.1.1 ROS 社区 ············ 3
   1.1.2 计算图 ·············· 5
   1.1.3 工作空间 ············ 10
1.2 ROS2 设计 ················ 13

第2章 初识 ROS2 ············ 15
2.1 首次了解 ROS2 ············· 15
2.2 开发第一个节点 ············ 20
2.3 分析 br2_basics 软件包 ····· 24
   2.3.1 控制迭代执行 ········ 25
   2.3.2 发布和订阅 ·········· 29
   2.3.3 启动器 ·············· 32
   2.3.4 参数 ················ 34

   2.3.5 执行器 ·············· 37
2.4 模拟机器人设置 ············ 38

第3章 第一个行为：用有限状态机避开障碍 ········ 44
3.1 感知和执行模型 ············ 45
3.2 计算图 ···················· 49
3.3 "碰撞与前进"在 C++ 中的实现 ···················· 50
   3.3.1 执行控制 ············ 50
   3.3.2 实现有限状态机 ······ 52
   3.3.3 运行代码 ············ 54
3.4 "碰撞与前进"在 Python 中的实现 ···················· 55
   3.4.1 执行控制 ············ 56
   3.4.2 实现有限状态机 ······ 57
   3.4.3 运行代码 ············ 58
建议的练习 ··················· 59

第4章 TF 子系统 ············ 60
4.1 使用 TF2 的障碍物检测器 ···· 65
4.2 计算图 ···················· 66
4.3 基础检测器 ················ 67

	4.3.1	障碍检测节点·············	68
	4.3.2	障碍物监控节点·········	70
	4.3.3	运行基本检测器·········	72
4.4	改进的检测器················		74
建议的练习·························			77

第 5 章 反应式行为·············· 78

5.1	使用 VFF 避免障碍物·········		78
	5.1.1	计算图·················	79
	5.1.2	软件包结构·············	80
	5.1.3	控制逻辑···············	81
	5.1.4	VFF 向量的计算·········	82
	5.1.5	使用视觉标记进行调试···	83
	5.1.6	运行 AvoidanceNode····	85
	5.1.7	开发过程中的测试·······	86
5.2	跟踪对象····················		92
	5.2.1	感知和执行模型·········	93
	5.2.2	计算图·················	97
	5.2.3	生命周期节点···········	98
	5.2.4	创建自定义消息·········	100
	5.2.5	跟踪实现···············	101
	5.2.6	执行跟踪器·············	110

建议的练习·························			112

第 6 章 用行为树对机器人行为进行编程············· 113

6.1	行为树······················		113
6.2	使用行为树实现"碰撞与前进"任务················		119
	6.2.1	使用 Groot 创建行为树··	121
	6.2.2	行为树节点实现·········	123
	6.2.3	运行行为树·············	126
	6.2.4	测试行为树节点·········	128
6.3	使用行为树进行巡逻·········		131
	6.3.1	Nav2 介绍···············	131
	6.3.2	设置 Nav2 参数·········	136
	6.3.3	计算图和行为树·········	140
	6.3.4	巡逻任务的实现·········	142
	6.3.5	运行巡逻任务···········	149
建议的练习·························			150

附录 源代码·················· 151

参考文献······················ 239

CHAPTER 1

第 1 章

引　言

机器人需要通过编程才能运行。即使机器人在机械设计方面非常出色，如果没有软件来处理传感器信息并向执行器发送命令以完成任务，那么它也没有任何用处。本章介绍了机器人编程所需的中间件，特别是 ROS2[8]，它也是本书使用的中间件。

首先，鉴于机器人软件非常复杂，不可能从零开始编写机器人程序。机器人必须在现实、动态和有时不可预测的环境中执行任务，同时还要处理各种不同型号和类型的传感器和执行器，这使得开发必要的驱动程序或适配新的硬件模块变得极为艰巨且容易失败。

中间件是介于操作系统和用户应用程序之间的软件层，用于支持应用程序的开发和运行。它通常不仅包括类库文件，还包括开发和监控工具及开发方法论。图 1.1 展示了机器人编程中的软件分层架构。

图 1.1　机器人编程中的软件分层架构

机器人编程中间件不仅提供驱动程序、库和方法，还提供开发、集成、执行和监控工具。在机器人学的发展历史中，出现了许多机器人编程中间件。然而，其中大部分都很难用于设计机器人,或从应用它们的实验室中进一步发展。尽管有一些著名的例子(如 Yarp[5]、Carmen[6]、Player/Stage[2] 等),但上一个十年里最成功的中间件毫无疑问是 ROS[7]。目前,它被认为是机器人编程领域的标准。其实,各种中间件在技术方面有很多相似之处:大多数都基于开源平台,并且提供分布式组件之间的通信机制、编译系统、监控工具等。但是,ROS 强大的开发者社区使其与众不同。此外,在该社区中,还有许多领先的企业、国际组织以及大学,它们提供了丰富的软件、驱动程序、文档和已解决的问题,这几乎涵盖了可能出现的任何问题。机器人学可以称之为"集成的艺术",ROS 提供了许多软件和相关工具来实现集成。

本书将提供 ROS2 编程项目所需的开发技能栈。由于我们将从头开始解释 ROS2 的概念、工具和方法,因此不需要读者有相关的经验。我们假设读者具备一些 Linux 操作系统和编程语言的基础。我们会使用 ROS2 官方支持的 C++ 和 Python,这两种语言也是机器人编程中最常用的编程语言。

1.1 ROS2 概述

ROS 的缩写含义是机器人操作系统(Robert Operating System)。它并不是用来取代 Linux 或 Windows 这类传统操作系统的,而是作为中间件来扩展系统开发机器人应用程序的能力⊖。数字"2"代表第二代的中间件。了解 ROS 第一代(常被称为 ROS1)的读者会发现 ROS2 中很多相似的概念,已经有相应的资源⊜可帮助 ROS1 程序员转向 ROS2 的开发。在本书中,我们假设读者对 ROS 是零基础的。现在有越来越多的人直接学习 ROS2,而不是先学习 ROS1,这种情况将变得越来越普遍。

另外,因为已经有一些优秀的官方 ROS2 教程,所以本书采取了不同的教学方法。本文将提供全面的描述和面向开发机器人应用程序的方法,使机器人能够实现"智能"功能,同时强调机器人软件开发过程中出现的一些基本问题。读者可以查看其他教程,以完善自己的培训体系并填补本书未涵盖的知识领域。

- ROS2 官方教程:https://docs.ros.org/en/humble/Tutorials.html。

本书起点是在一台配备 AMD64 位架构的个人笔记本电脑或台式电脑上安装 Linux Ubuntu 22.04 LTS 系统。由于 ROS2 是按发行版组织的,因此 Linux 发行版的选择很重要。发行版是一组经过验证且能够协同工作的库、工具和应用程序的集合。每个发行版都有一个特定的名称,并与 Ubuntu 的一个版本相关联。发行版中的软件也保证与系统上的软件版本正确配合使用。虽然可以使用其他 Linux 发行版(如 Ubuntu、Fedora、Red Hat 等),但本文将以 Ubuntu 22.04 为参考。虽然 ROS2 也

⊖ ROS 并未被百分百定义为中间件,它实际是中间件、框架和元操作系统的混合体。
⊜ https://github.com/fmrico/ros_to_ros2_talk_examples。

可以在 Windows 和 Mac 上运行，但本书会重点介绍 Linux 的开发环境。我们将使用 ROS2 Humble Hawksbill 版本，它与 Ubuntu 22.04 兼容。

在本书中，我们将从三个不同但相互补充的维度来深入了解 ROS2：

- **ROS 社区**：ROS 社区是开发此类机器人应用时的一个基本要素。除了提供技术文档外，它还有一个庞大的开发者社区。社区成员通过公共代码仓库分享自己的应用程序和实用工具，其他开发者也可以参与。因此，社区中可能已经有你所需的功能或解决方案。
- **计算图**：计算图是运行中的 ROS2 应用程序的可视化表示。它由节点和弧组成。节点是 ROS2 中的主要计算单元，可以通过几种不同的通信范式与其它节点协同工作，以构建一个 ROS2 应用程序。此维度还涉及监控工具，这些工具也是插入到计算图中的节点。
- **工作空间**：工作空间是指安装在机器人或计算机上的软件集合，以及用户所开发的程序。不同于计算图的动态特性，工作空间是静态的。此维度还探讨了用于构建计算图元素的开发工具。

1.1.1 ROS 社区

ROS 社区是 ROS2 蓬勃发展的支柱。开源机器人基金会⊖的成立在很大程度上推动了用户与开发者社区的发展。ROS2 不仅仅是一个机器人编程的中间件，还是一种开发方法论，它建立了软件的交付机制，并提供了一系列资源供 ROS 社区成员使用。

ROS2 属于开源软件。这意味着它是一种以开源许可证形式发布的软件，用户具有使用、研究、修改和再分发的权利。许多开源许可证对这类软件的某些自由做出了一定的调控，但基本上我们可以假设拥有这些权利。ROS2 软件包最常见的许可证是 Apache 2 和 BSD，开发者也可以自由选择其他许可证。

ROS2 采用联邦模型来组织软件的交付，并提供了相关的保障技术机制。每个开发者、公司或机构都能自由地开发自己的软件，并负责管理。同时，小型项目通常会围绕它自身形成一个社区，该社区可以对这个小型项目的软件发布事项进行决策。这些实体机构可创建 ROS2 软件包，并将其提供给公共仓库或作为 ROS2 发行版二进制文件的一部分。没有人会强制这些实体机构迁移其软件到 ROS2 的新版本。即便如此，许多重要且受欢迎的软件包的持续性发展足以确保它们的延续性。

这种开发模式的重要性在于促进 ROS 社区的发展。从实用的角度来看，这对于机器人编程中间件的成功至关重要。这种中间件的一个期望特性是对许多传感器和执行器提供支持。目前，许多这些组件的制造商都支持其 ROS2 驱动程序，因为它们知道有许多潜在客户，并且有许多开发人员在购买之前会检查特定组件是否在

⊖ https://www.openrobotics.org。

ROS2 中得到支持。此外，这些制造商通常会在公共的代码仓库中开发这些软件，从而构建自己的用户社区以报告缺陷甚至提交对应的补丁程序。如果你希望大范围推广你的机器人库或工具，那么支持 ROS2 可能是正确的选择。

ROS2 中的软件包按发行版组织。一个 ROS2 发行版由众多协同良好的软件包组成。通常情况下，ROS2 发行版与特定版本的操作系统绑定在一起。ROS2 使用 Ubuntu Linux 版本作为参考，以确保其稳定性，否则不同软件包和库的版本依赖关系问题会使 ROS2 变得一团糟。当实体组织发布特定软件时，会针对其对应的某个发行版发布该软件。每个发行版通常会维护多个开发分支。

截至 2023 年 1 月，ROS2 一共推出了 8 个发行版本，见表 1.1。每个发行版都有一个以字母表顺序递增的名称和一个不同的徽标（还有不同的 T 恤模型），见图 1.2。Rolling Ridley 比较特殊，它既是作为 ROS2 未来稳定版的一个临时区域，也是最新开发版本的集合地。

表 1.1　ROS2 一共推出了 8 个发行版本

发行版名称	发行日期	终止服务日期	Ubuntu 版本
Humble Hawksbill	2022 年 5 月 23 日	2027 年 5 月	Ubuntu 22.04
Galactic Geochelone	2021 年 5 月 23 日	2022 年 11 月	Ubuntu 20.04
Foxy Fitzroy	2020 年 6 月 5 日	2023 年 5 月（长期支持）	Ubuntu 20.04
Eloquent Elusor	2019 年 11 月 22 日	2020 年 11 月	Ubuntu 18.04
Dashing Diademata	2019 年 5 月 31 日	2021 年 5 月（长期支持）	Ubuntu 18.04
Crystal Clemmys	2018 年 12 月 14 日	2019 年 12 月	Ubuntu 18.04
Bouncy Bolson	2018 年 7 月 2 日	2019 年 7 月	Ubuntu 16.04
Ardent Apalone	2017 年 12 月 8 日	2018 年 12 月	Ubuntu 16.04

图 1.2　截至 2023 年 1 月发布的 ROS2 发行版

如果你想将软件贡献到 ROS 发行版中，欢迎访问 rosdistro 代码仓库（https://github.com/ros/rosdistro），下面是一些可能对你有用的链接：

- 发布你的软件包：https://docs.ros.org/en/rolling/How-To-Guides/Releasing/Releasing-a-Package.html。

开源机器人基金会为社区提供了许多资源，例如：

- ROS 官方页面：http://ros.org。
- ROS2 文档页面：https://docs.ros.org。每个发行版都有其相应的文档。例如，在 https://docs.ros.org/en/humble 中，你可以找到安装指南、教程等。
- ROS 问答（https://answers.ros.org）是一个提供解决 ROS 相关问题和疑惑的地方。
- ROS 讨论（https://discourse.ros.org）是一个专门为 ROS 社区而设的在线讨论论坛。你可以通过 ROS 讨论及时跟进社区最新动态、查看发布公告或讨论设计问题。此外，ROS 讨论还设有多种语言的 ROS2 用户组。

1.1.2 计算图

在计算图这个维度中，我们将分析机器人软件在执行过程中的运行情况。这个视角将为我们提供一个目标的概念，它将有助于我们更容易理解接下来的内容。

一个计算图包含了相互通信的 ROS2 节点，以便机器人能够执行一些任务。节点作为 ROS2 中主要的执行元素，承载了应用程序的逻辑。

ROS2 广泛地使用面向对象编程。一般来说，无论是用 C++ 还是 Python 编写，一个节点都是节点类的一个实例对象。

一个节点可以访问计算图，并通过以下三种模式与其它节点进行通信：

- **发布/订阅**：一种异步通信方式，其中 N 个节点发布消息到一个消息主题，并到达其 M 个订阅者。消息主题就像一个接受同一类型消息的通信渠道。这种通信方式在 ROS2 中最为常见。一个非常典型的案例是包含摄像头驱动程序的节点发布图像到一个主题。需要接受来自摄像头的图像以完成其功能的系统中的所有节点都将订阅该主题。
- **服务**：一种同步通信方式[⊖]，其中一个节点请求另一个节点并期待立即响应。这种通信方式通常需要即时回应，以免影响调用服务的节点的控制周期。例如请求地图映射服务来重置地图，并通过响应获知调用是否成功。
- **动作**：一种异步通信方式，其中一个节点向另一个节点发送请求。这些请求通常需要一段时间才能完成，并且调用节点可能会定期收到反馈或获知请求已成功完成或出现了一些错误的通知。导航请求就是这种通信方式的一个例

⊖ 这种通信方式在 ROS1 中是同步的，但在 ROS2 中不建议采用同步服务客户端（https://docs.ros.org/en/humble/How-To-Guides/Sync-Vs-Async.html）。

子。这个请求可能需要很长时间来完成，因此请求机器人导航的节点不应该在完成过程中被阻塞等待。

在计算图中，一个节点的作用是执行处理或控制。因此，它们被视为具备执行模型选择权的主动元素：

- **周期执行**：在控制软件中，一个节点以特定的频率执行其控制周期是很常见的。这种方法允许控制节点所需的计算资源数量，并且输出流保持不变。例如，一个节点根据执行器的状态以 20Hz 的频率计算运动指令。
- **事件驱动执行**：这些节点的执行是由某些事件发生的频率决定的，如消息到达此节点的频率。例如，某一节点每接收到一张图像就对其进行检测并生成输出。输出发生的频率取决于图像到达的频率。如果没有图像到达，就不会产生输出。

接下来，我们将展示几个计算图的示例。图 1.3 呈现的是这些示例中使用到的符号描述。

图 1.3　计算图中使用到的符号描述

第一个计算图，如图 1.4 所示，是一个与 Kobuki 机器人交互的简单程序示例，Kobuki 是一款基于 Roomba 的小型移动机器人。

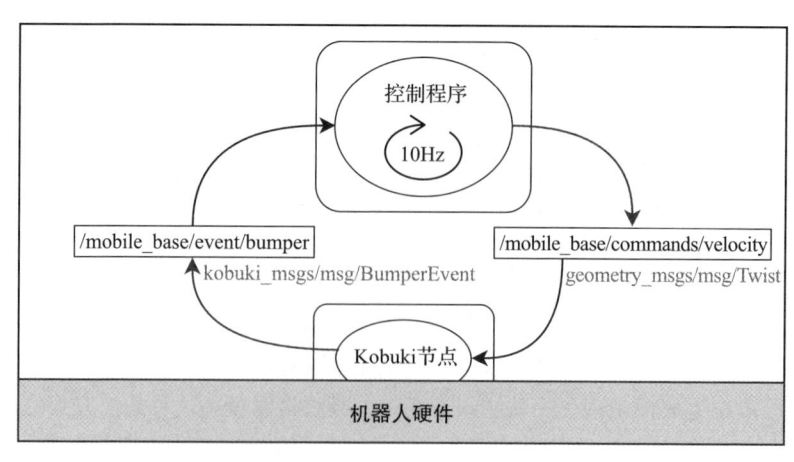

图 1.4　一个针对 Kobuki 机器人的简单控制的计算图。控制应用程序从其订阅的碰撞传感器的信息中计算出速度，并将其发布出去

Kobuki 机器人驱动程序是一个节点，通过使用原生驱动程序与机器人硬件进行通信。它的功能通过各种不同的消息主题暴露给用户。在本例中，我们仅展示了两个消息主题：

- /mobile_base/event/bumper：当机器人 Kobuki 的碰撞传感器状态发生变化时，会发布一个 kobuki_msgs/msg/BumperEvent 消息到这个消息主题下，所有对这个传感器检测碰撞感兴趣的系统节点都会订阅这个主题。
- /mobile_base/commands/velocity：Kobuki 驱动程序订阅的这个主题用于调整其速度。如果在一秒钟内没有收到任何指令，那么它将停止。该消息主题属于 geometry_msgs/msg/Twist 类型。几乎所有 ROS2 移动机器人都会接收这些类型的消息以控制它们的速度。

> **深入了解：ROS2 中的命名**
>
> 在 ROS2 中，资源的命名遵循类似 Unix 文件系统的约定。当创建一个资源（例如发布者）时，我们可以指定其相对名称、绝对名称（以"/"开头）或私有名称（以"~"开头）。此外，我们还可以定义命名空间，目的是通过将工作空间的名称作为名称的第一个组成部分，将资源与其他命名空间隔离开来。命名空间在多机器人应用程序中非常有用。让我们看一个基于节点名称和命名空间生成主题名称的示例：
>
> | 名称 | 结果
（节点：my_node/ 命名空间：无） | 结果
（节点：my_node/ 命名空间：my_ns） |
> |---|---|---|
> | my_topic | /my_topic | /my_ns/my_topic |
> | /my_topic | /my_topic | /my_topic |
> | ~my_topic | /my_node/my_topic | /my_ns/my_node/my_topic |
>
> 延伸阅读：
> - http://wiki.ros.org/Names

这个节点在一个单独的进程中运行。计算图显示了另一个订阅碰撞传感器主题的进程，并根据接收到的信息发布机器人应该移动的速度。我们已经设置了节点的执行频率，以表示它用每秒 10 次的频率进行控制决策，不论是否接收到关于碰撞传感器状态的消息。

这个计算图包括两个节点和两个消息主题，每个消息主题都有其对应的发布/订阅连接。

让我们进一步改进增强这个机器人和应用程序，即加入一个激光传感器和一个

3D 相机（也称为 RGBD 相机）。对于每个传感器，都需要有一个节点来接入传感器并通过 ROS2 接口呈现它。正如先前所述，从传感器上发布数据是让数据在计算图中变得可用的最快捷的方式。

应用程序现在可使机器人根据从 RGBD 相机检测到的 3D 图像中的人或物体的位置信息进行移动，移动过程中激光传感器可避免碰撞。图 1.5 是一个改进的计算图。

- 控制节点以 20Hz 的频率运行，向机器人节点发送控制命令。机器人节点订阅此主题 /scan 以检查周围的障碍物。
- 该进程包含两个节点，分别用于侦测人体和物体。这两个节点都需要从相机获取图像和深度信息，以确定侦测对象的位置。每次检测结果都会以符合 3D 侦测标准的消息类型发布到两个不同的消息主题中。
- 控制节点订阅这些消息主题，并根据侦测结果执行相应任务。

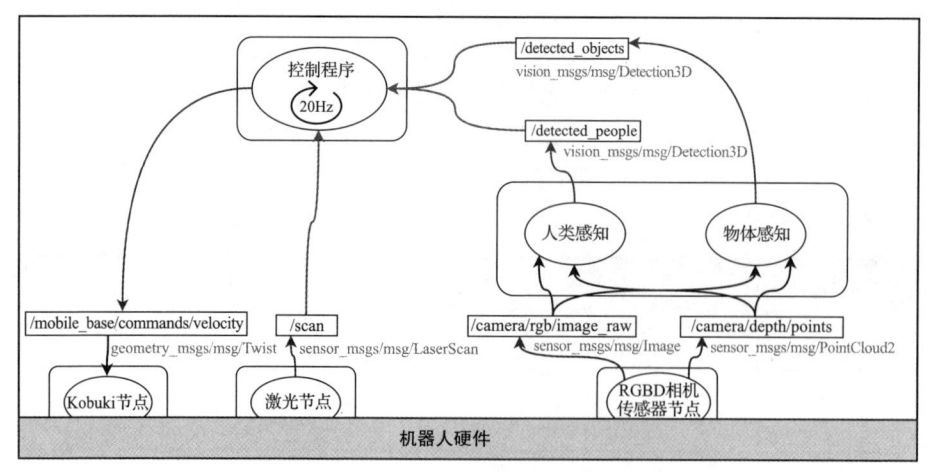

图 1.5 一个改进的计算图，该图中的控制应用程序使用了激光数据和从机器人的 RGBD 相机获取的经过预处理的信息（人和物体）

使用上一个示例中 Tiago 机器人，假设只有一个节点提供了其功能。这里使用两个订阅者（用于移动 Tiago 的底座的速度命令和用于移动 Tiago 的颈部的轨迹命令）和两个发布者（激光信息和来自 RGBD 相机的 3D 图像）。

Tiago 机器人应用程序分为两个子系统（行为子系统和导航子系统），每个子系统运行在一个不同的进程中，每个进程包含对应子系统的节点（我们省略了每个子系统中主题的细节）。该应用程序的计算图如图 1.6 所示。

- **行为子系统：** 它由两个节点组成，这两个节点相互协作以生成机器人的行为。

其中一个节点是协调器（行为协调器），另一个节点是实现主动视觉系统的头控制器。协调器确定视线位置和机器人在地图上应该访问的点。
- **导航子系统**：它由多个节点组成。导航管理器协调规划器（负责创建从机器人的位置到目的地的路线）和控制器（使机器人遵循创建的路线）的工作。规划器需要地图（由加载环境地图的节点提供），还需要机器人的位置（由计算位置的节点计算得出）。
- 行为子系统和导航子系统之间的通信是通过 ROS2 的动作模式（action）来完成的。导航行为设置一个目标，并在完成时收到通知。它还会周期性地接收到达目的地的进度。动作模式（action）也用于协调导航系统中的规划器和控制器。

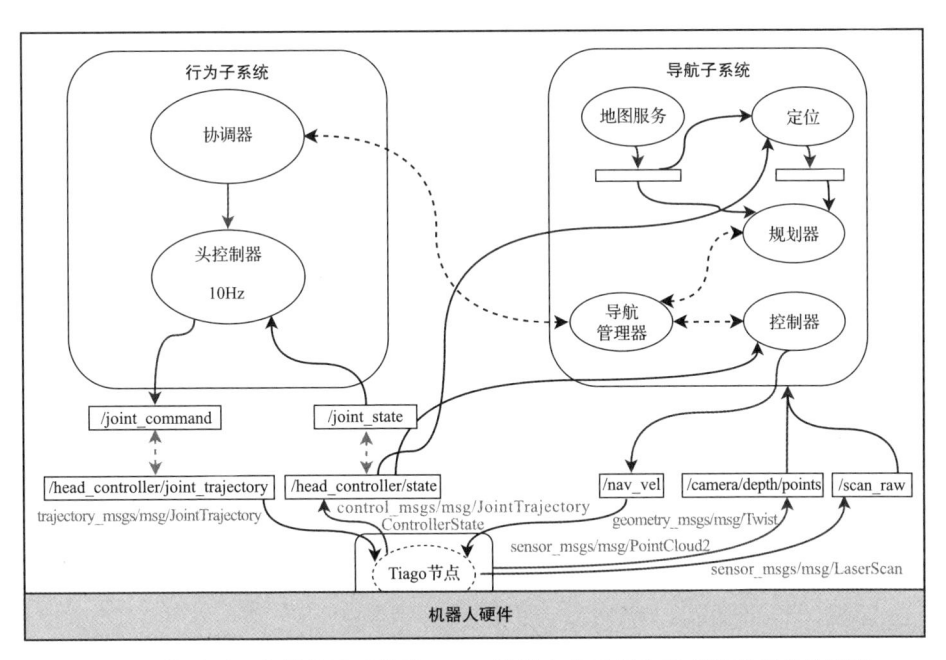

图 1.6　使用了一个导航子系统的 Tiago 机器人基于行为的应用程序的计算图

在本小节中，我们展示了几个计算图。每次在 ROS2 中实现应用程序时，都需要先设计一个计算图，再确定所需的节点及其交互方式。我们必须决定节点是按特定频率执行，还是由某个事件（请求或消息）触发其执行。我们可以自行开发所有的节点，也可以在计算图中包含第三方开发的节点。

尽管我们可以在应用程序中定义新的消息类型，但 ROS2 已经定义了一组标准消息类型，以便于不同开发者开发的节点之间的交互。例如，为图像定义一种新的消息类型是不明智的，因为有很多第三方软件、处理和监控工具会使用和生成被视为图像标准的消息类型（sensor_msgs/msg/Image）。因此，尽可能采用已有的标准化消息类型。

1.1.3 工作空间

工作空间（workspace）为我们提供了一个静态视角，用于理解 ROS2 软件的构成。通过工作空间，我们可以了解 ROS2 软件的安装位置、组织结构，以及启动计算图所需的工具和指令。

在工作空间的视角下，构成 ROS2 软件的基本元素是软件包（package）。软件包是指具有共同目标的可执行文件、库或消息定义。通常，软件包依赖其他软件包来运行或构建。

工作空间是一个包含软件包的目录。因此，工作空间本身也是 ROS 软件的一个重要组成部分，需要激活该工作空间，才能使用其中的内容。

可以同时激活多个工作空间，且激活过程是累积的。我们可以激活一个初始工作空间，称为底层。之后，可以激活另一个工作空间，称为叠加层，因为它覆盖了之前的底层工作空间。在底层中应满足叠加层中软件包的依赖关系。如果叠加层中的包已存在于底层中，则叠加层中的软件包将隐藏底层中的相应软件包。

通常，包含基本 ROS2 安装的工作空间会被立即激活。这是 ROS2 系统中最常见的底层。然后，用户开发自己软件包的工作空间被激活。

软件包可以通过源代码安装，也可以通过系统的安装机制进行安装。在本书参考的 Ubuntu Linux 22.04 系统中，使用像 apt 这样的工具通过 deb 包进行安装。每个 ROS2 软件包都被打包成一个 deb 包。一个发行版中的 deb 包的名称很容易识别，因为它们的名称以 ros-<distro>-<ros 2 package name> 开头。为了访问这些包，需要配置 APT ROS2 仓库：

```
$ sudo apt update && sudo apt install curl gnupg2 lsb-release
$ sudo curl -sSL https://raw.githubusercontent.com/ros/rosdistro/master/ros.key -o /usr/share/keyrings/ros-archive-keyring.gpg
$ echo "deb [arch=$(dpkg --print-architecture) signed-by=/usr/share/keyrings/ros-archive-keyring.gpg] http://packages.ros.org/ros2/ubuntu $(source /etc/os-release && echo $UBUNTU_CODENAME) main" | sudo tee /etc/apt/sources.list.d/ros2.list > /dev/null

$ sudo apt-get update
```

当然，deb 软件包的安装依赖 ROS2 软件包的依赖项。下面的命令显示可供安装的 ROS2 软件包：

```
$ apt-cache search ros-humble
```

基本上，ROS2 Humble 只需要输入以下内容即可安装：

```
$ sudo apt update
$ sudo apt install ros-humble-desktop
```

所有通过 apt 安装的 ROS2 软件都位于 /opt/ros/humble。在 Ubuntu 22.04 系统上，也可以安装 ROS2 Rolling 版本或 ROS Iron 版本。如果两者都安装了，则分别位于 /opt/ros/rolling 和 /opt/ros/iron。我们甚至可以通过在其他位置编译其源代码来安装这些 ROS 发行版之一。由于不推荐混合使用 ROS 发行版（除非你知道自己在做什么），因此安装发行版并不会自动"激活"它。激活是通过在终端执行以下命令完成的：

```
$ source /opt/ros/humble/setup.bash
```

此命令会激活位于 /opt/ros/humble 的软件。通常将此行添加到 $HOME/.bashrc 文件中，这样在打开终端时它会默认被激活：

```
$ echo "source /opt/ros/humble/setup.bash" >> ~/.bashrc
```

安装和配置 rosdep[⊖] 工具也很方便。这个工具在一组源代码包中查找不满足的依赖项，并将它们作为 deb 包进行安装。我们只需要在安装后运行一次这些命令：

```
$ sudo rosdep init
$ rosdep update
```

通常用户会在 $HOME 目录中创建一个目录，其中包含正在开发的包的源代码。让我们仅通过创建一个包含 src 目录的目录来创建一个工作空间。然后添加本书中将使用的示例软件包。

```
$ cd
$ mkdir -p bookros2_ws/src
$ cd bookros2_ws/src
$ git clone -b humble-devel https://github.com/fmrico/book_ros2.git
```

如果我们查看在 src 下添加的内容，那么我们将能够看到目录的集合。那些在根目录中有 package.xml 文件的目录就是软件包的位置。

在这个工作空间中，有许多软件包依赖 ROS2 Humble 发行版中没有的其他软件包。为了将这些软件包的源代码添加到工作空间中，我们将使用 vcstool[⊖]：

⊖ http://wiki.ros.org/rosdep。

⊖ https://github.com/dirk-thomas/vcstool。

```
$ cd ~/bookros2_ws/src
$ vcs import . < book_ros2/third_parties.repos
```

vcs 命令从 .repos 文件读取仓库列表，并将它们复制到指定目录。在构建之前，让我们使用 rosdep 安装构建整个工作空间所缺少的任何软件包：

```
$ cd ~/bookros2_ws
$ rosdep install --from-paths src --ignore-src -r -y
```

一旦带有依赖项的示例软件包的源代码位于工作空间中，就可以始终使用 colcon[⊖]命令从其根目录开始构建工作空间：

```
$ cd ~/bookros2_ws
$ colcon build --symlink-install
```

检查一下，在工作空间根目录下是否创建了以下三个文件夹：
- build：该文件夹包括编译的中间文件、测试文件和临时文件。
- install：该文件夹包含编译结果以及执行需要的所有文件（具体的配置文件、节点启动脚本、映射等）。在使用 --symlink-install 选项构建工作空间时，会创建到原始位置（在 src 或 build 中）的符号链接，而不是复制文件。通过这种方式，我们可以节省空间，并且可以直接在 src 中修改某些配置文件。
- log：该文件夹包含编译或测试过程生成的日志。

> **深入了解：colcon**
>
> colcon（集体构建）是一个用于构建、测试和使用多个软件包的命令行工具。通过 colcon，你可以编译 ROS1、ROS2 甚至是 cmake 包。它自动化了构建过程并设置了使用这些软件包的环境。
>
> 延伸阅读：
> - https://colcon.readthedocs.io
> - https://vimeopro.com/osrfoundation/roscon-2019/video/379127725

要清理或重置工作空间，只需要删除上述三个文件夹。新的编译将会重新生成它们。

为了使用工作空间的内容，请将它激活为叠加层，其方式与底层的激活方式类似：

⊖ https://colcon.readthedocs.io/en/released/index.html。

```
$ source ~/bookros2_ws/install/setup.bash
```

通常在 $HOME/.bashrc 文件中包含以下行，这样在打开终端时它们会被默认激活：

```
$ echo "source ~/bookros2_ws/install/setup.bash" >> ~/.bashrc
```

1.2 ROS2 设计

图 1.7 展示了 ROS2 的分层设计。用户开发的节点下方的层为程序员提供了一个与 ROS2 交互的 API。通过 C++ 实现的节点和程序的包使用 C++ 客户端库 rclcpp，而 Python 中的包则使用 rclpy。

图 1.7　ROS2 的分层设计

rclcpp 和 rclpy 并不是完全独立的 ROS2 实现。如果是这样的话，用 Python 编写的节点可能与用 C++ 编写的节点具有不同的行为。rclcpp 和 rclpy 都使用了 rcl，它实现了所有 ROS2 元素的基本功能。rclcpp 和 rclpy 将这些功能适应到每种语言的特性上，同时还会进行其他语言特性所需的更改，例如线程模型。

任何其他语言（如 Rust[一]、Go[二]、Java[三]、.NET[四]等）的客户端库都应构建在 rcl

[一] https://github.com/ros2-rust/ros2_rust。
[二] https://github.com/tiiuae/rclgo。
[三] https://github.com/ros2-java/ros2_java。
[四] https://github.com/ros2-dotnet/ros2_dotnet。

之上。

  rcl 是 ROS2 的核心。没有人直接使用它来编写程序。如果用户想要使用 C 语言编写节点，则可以使用 rclc 的 ROS2 的 C 客户端库。虽然它使用与 rcl 相同的 C 语言编写，但它封装了更高级的功能，比直接使用 rcl 更加简单方便。

  ROS2 的一个重要的组件是通信层。ROS2 是一个分布式系统，其计算图中的节点可以分布在多台机器上。即使所有软件都运行在单个机器人上，节点也可以在操作员的个人电脑上运行来监控和控制机器人的操作。

  ROS2 选择了数据分发服务（Data Distribution Service，DDS）[一]作为其通信层，这是一种基于 UDP 协议实现的下一代通信中间件，它允许进程之间以实时性、安全性和自定义服务质量的方式交换信息。DDS 提供了发布/订阅通信模式，提供了一种在不需要集中式服务的情况下发现发布者和订阅者的机制。这种发现机制使用了多播（multicast），尽管随后的连接是单播的（unicast）。

  市面上有数家 DDS 厂商，包括 FastDDS[二]、CycloneDDS[三]或 RTI® Connext[四]。这些厂商全部或部分实现了由 OMG®定义的 DDS 标准。ROS2 可以使用所有这些 DDS 实现。但是，当我们在低延迟、大量数据或资源有限等场景下追求更高性能时，就需要根据上述 DDS 实现的差异选择符合相应场景标准的实现方式。

  这些 DDS 具体实现的 API 接口各不相同。因此，为了简化 rcl 层，定义了一个名为 rmw 的底层抽象实现，它为 rcl 程序员提供了一个统一的 API 接口来访问每个支持 DDS 实现的功能。选择要使用的 DDS 时，只需要修改一个环境变量 RMW_IMPLEMENTATION 即可。

  Humble 发行版中 DDS 的官方实现是 FastDDS，而在 Galactic 发行版中，DDS 的官方实现是 CycloneDDS，由此导致了供应商间的竞争。希望竞争的主要受益者将是 ROS2 社区。

[一] https://www.omg.org/omg-dds-portal。
[二] https://github.com/eProsima/Fast-DDS。
[三] https://github.com/eclipse-cyclonedds/cyclonedds。
[四] https://www.rti.com/products/dds-standard。
[五] https://www.omg.org。

CHAPTER 2

第 2 章

初识 ROS2

上一章介绍了 ROS2 的基本理论概念以及如何安装 ROS2。在本章中,我们将开始使用 ROS2 进行练习,并学习 ROS2 的第一个概念。

2.1 首次了解 ROS2

安装完 ROS2 后,将所需的环境变量添加到~/.bashrc 中,以确保底层(/opt/ros/humble)和叠加层(~/bookros2_ws)都已经被激活。可以输入以下命令进行检查:

```
$ ros2
usage: ros2 [-h] Call 'ros2 <command> -h' for more detailed usage. ...
ros2 is an extensible command-line tool for ROS2.
...
```

如果底层依赖文件被激活,那么将会找到此命令。

ros2 是 ROS2 中的主要快捷命令,可以通过其与 ROS2 系统交互以获取信息或执行指令。

```
ros2 <command> <verb> [<params>|<option>]*
```

输入以下命令可以获取可用包的列表:

```
$ ros2 pkg list
ackermann_msgs
```

```
action_msgs
action_tutorials_cpp
...
```

这个示例中，pkg 包管理器命令负责 ROS2 的软件包的管理。list 命令参数会显示底层（全局）或任何叠加层（用户自定义）中的依赖的软件包列表。

> **深入了解：ros2cli**
>
> ros2cli 是 ROS2 命令行工具。它是模块化且可扩展的，因此可以通过 roscli 添加更多功能。目前支持的标准操作是：
>
> | | | | |
> |---|---|---|---|
> | action | extension_points | node | test |
> | bag | extensions | param | topic |
> | component | interface | pkg | wtf |
> | launch | run | daemon | lifecycle |
> | security | doctor | multicast | service |
>
> 延伸阅读：
> - https://github.com/ros2/ros2cli
> - https://github.com/ubuntu-robotics/ros2_cheats_sheet/blob/master/cli/cli_cheats_sheet.pdf

ros2 命令支持 Tab 键自动补全功能。可以输入 ros2，然后按两次 Tab 键以查看可以用的命令，也可以使用 Tab 键查找命令的参数。

另外，它也可以获取有关特定包的信息，例如，要从 demo_nodes_cpp 包中获取可执行程序：

```
$ ros2 pkg executables demo_nodes_cpp
demo_nodes_cpp add_two_ints_client
demo_nodes_cpp add_two_ints_client_async
demo_nodes_cpp add_two_ints_server
demo_nodes_cpp allocator_tutorial
...
demo_nodes_cpp talker
...
```

使用 run 命令执行其中一个命令，这需要两个参数：可执行文件所在的软件包的名称和可执行程序的名称。这个软件包的名称表示它包含的所有程序都是用 C++ 编写的。

```
$ ros2 run demo_nodes_cpp talker
[INFO] [1643218362.316869744] [talker]: Publishing: 'Hello World: 1'
[INFO] [1643218363.316915225] [talker]: Publishing: 'Hello World: 2'
[INFO] [1643218364.316907053] [talker]: Publishing: 'Hello World: 3'
...
```

正如所见，当使用软件包的名称和可执行程序的名称来指定要执行的程序时，无须知道这些程序的确切位置，也不需要在任何特定的位置执行它们。

如果一切顺利，终端中就会出现带有计数器的"Hello world"消息。保持此命令运行并打开另一个终端以查看此可执行文件正在执行的操作。在ROS2中，常常会同时打开多个终端，所以在屏幕上合理布局很有必要。talker节点的计算图如图2.1所示。

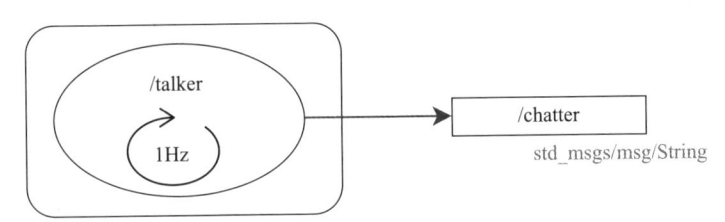

图 2.1　talker 节点的计算图

使用 node 命令及其 list 参数检查当前正在运行的节点，在另一个终端中执行如下命令：

```
$ ros2 node list
/talker
```

这个命令确认只有一个名为 /talker 的节点。在 ROS2 中的资源名称，例如节点，其格式与 Linux 系统中的文件类似。从 /root 根目录开始，斜杠（/）用于分隔名称的不同部分。

/talker 节点不仅通过终端打印信息消息，还会向消息主题发布消息。

当 /talker 节点运行时，可以使用 topic 命令和 list 参数来检查系统中有哪些消息主题。

```
$ ros2 topic list
/chatter
/parameter_events
/rosout
```

通过 topic list 命令查看多个消息主题，其中包括发布 /talker 节点的 /chatter 消息主题。可以使用 node 命令的 info 参数来获取更多信息：

```
$ ros2 node info /talker
/talker
    Subscribers:
        /parameter_events: rcl_interfaces/msg/ParameterEvent
    Publishers:
        /chatter: std_msgs/msg/String
        /parameter_events: rcl_interfaces/msg/ParameterEvent
        /rosout: rcl_interfaces/msg/Log
    Service Servers:
...
```

输出结果显示有多个发布者，这与上一条命令显示的消息主题一致，因为系统中没有其他节点。

正如我们已经说过的，每个消息主题仅支持一种类型的消息。前面的命令已经显示了类型，也可以通过发送 topic 命令询问指定消息主题的信息来验证：

```
$ ros2 topic info /chatter
Type: std_msgs/msg/String
Publisher count: 1
Subscription count: 0
```

按照惯例，消息在包中定义，并以 _msgs 结尾。std_msgs/msg/String 是在 std_msgs 包中定义的字符串消息。要检查系统中的有效消息，请使用 interface 命令及 list 参数。

```
$ ros2 interface list
Messages:
    ackermann_msgs/msg/AckermannDrive
    ackermann_msgs/msg/AckermannDriveStamped
    ...
    visualization_msgs/msg/MenuEntry
Services:
    action_msgs/srv/CancelGoal
    ...
    visualization_msgs/srv/GetInteractiveMarkers
Actions:
    action_tutorials_interfaces/action/Fibonacci
    ...
```

上面的结果显示了所有类型的接口，通过这些接口，节点可以在 ROS2 中进行通信。添加 -m 选项，可以仅过滤显示消息。注意，除了消息之外还有其他更多的接口。服务和动作也可以用 ros2 interface 进行检查。

检查消息格式以获取消息中包含的字段及其类型如下：

```
$ ros2 interface show std_msgs/msg/String
... comments
string data
```

此消息格式只有一个名为 data 的字段，其类型为字符串。

> **深入了解：interface**
>
> 一个消息由多个字段组成。每个字段都有一个不同的类型，它可以是基本类型（布尔值、字符串、64 位浮点数）或消息类型。这样通常可以通过组合简单的消息来创建更复杂的消息。
>
> Stamped 消息就是一个例子。一系列的消息，其名称以 Stamped 结尾，会在现有的消息中增加一个头部。比较以下两个消息的差异：
>
> geometry_msgs/msg/Point
> geometry_msgs/msg/PointStamped
>
> 延伸阅读：
> - https://docs.ros.org/en/humble/Concepts/About-ROS-Inter-faces.html

检查当前正在发布的消息（/talker 应仍在另一个终端运行）。通过输入以下命令查看该消息主题上发布的消息：

```
$ ros2 topic echo /chatter
data: 'Hello World: 1578'
---
data: 'Hello World: 1579'
...
```

接下来，执行一个包含订阅 /chatter 消息主题的节点的程序，并将其接收到的消息显示在屏幕上。为了在不停止包含 /talker 节点的程序的情况下执行它，我们运行位于同名程序中的 /listener 节点。尽管在 demo_nodes_cpp 包中有一个 listener 节点，但为了多样化起见，从 demo_nodes_py 包中采用 Python 实现运行另一个 listener 节点。

```
$ ros2 run demo_nodes_py listener
[INFO] [1643220136.232617223] [listener]: I heard: [Hello World: 1670]
[INFO] [1643220137.197551366] [listener]: I heard: [Hello World: 1671]
[INFO] [1643220138.198640098] [listener]: I heard: [Hello World: 1672]
...
```

现在计算图由两个节点组成，它们通过 /chatter 消息主题进行通信。运行

Listener 节点的计算图如图 2.2 所示。

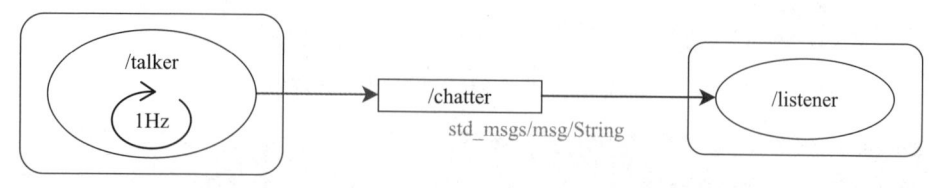

图 2.2　Listener 节点的计算图

也可以通过运行 rqt_graph 工具（如图 2.3 所示）来可视化计算图，该工具位于 rqt_graph 包中。

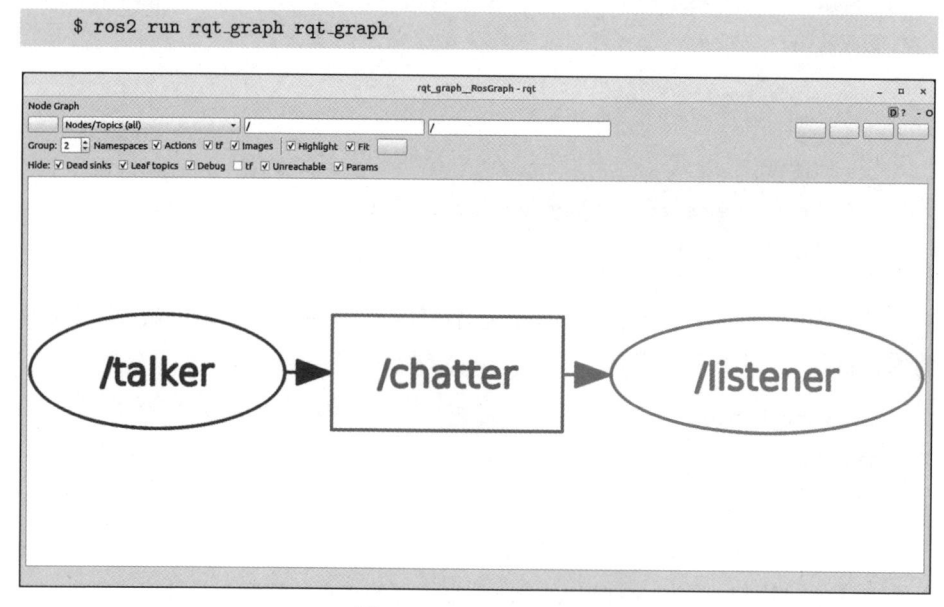

```
$ ros2 run rqt_graph rqt_graph
```

图 2.3　rqt_graph 工具

我想你已经理解了 ros2 的基础命令，现在可以在终端（命令行）中使用 "Ctrl+C" 来终止所有运行中的程序。

2.2　开发第一个节点

到目前为止，我们只使用了 ROS2 基础安装部分包的软件。在本节中，我们将创建一个包来开发第一个节点。

新包将在叠加层中创建（cd ~/bookros2_ws），以练习从头开始创建包。

所有软件包都必须位于 src 目录中。这次，我们使用 ros2 命令、pkg 命令以及 create 参数创建。在 ROS2 包中，使用这个命令的时候需要声明它们所依赖的其他包，无论是在这个工作空间还是其他工作空间，以便编译工具知道它们被构建的顺序。切换到 src 目录并运行：

```
$ cd ~/bookros2_ws/src
$ ros2 pkg create my_package --dependencies rclcpp std_msgs
```

这个命令构建了基础软件包的骨架，其中包含一些空目录来承载程序和库的源文件。一个目录只要含有一个 package.xml 的文件就会被 ROS2 识别出包含一个包。--dependencies 选项允许你添加这个包的依赖项。现在，我们将使用 rclcpp 指令，这个指令属于 C++ 客户端库。

```
package.xml
<?xml version="1.0"?>
<?xml-model href="http://download.ros.org/schema/package_format3.xsd"
    schematypens="http://www.w3.org/2001/XMLSchema"?>
<package format="3">
  <name>my_package</name>
  <version>0.0.0</version>
  <description>TODO: Package description</description>
  <maintainer email="john.doe@evilrobot.com">johndoe</maintainer>
  <license>TODO: License declaration</license>

  <buildtool_depend>ament_cmake</buildtool_depend>

  <depend>rclcpp</depend>
  <depend>std_msgs</depend>

  <test_depend>ament_lint_auto</test_depend>
  <test_depend>ament_lint_common</test_depend>

  <export>
    <build_type>ament_cmake</build_type>
  </export>
</package>
```

尽管 ros2 pkg create 是创建新包的一个好的起点，但在实践中，通常是通过复制现有的包来创建的，先更改软件包的名称再修改具体内容，这样的方式更方便。

由于示例是一个 C++ 包，因此我们明确指出它依赖 rclcpp，在其根目录中还创建了一个 CMakeLists.txt 文件，该文件建立了编译它的规则。我们将在添加一些要编译的内容后分析其内容。

首先，在 ROS2 中创建一个尽可能简单的程序，命名为 src/simple.cpp。以下包含了包结构和 src/simple.cpp 的源代码：

```
Package my_package
my_package/
├── CMakeLists.txt
├── include
│   └── my_package
├── package.xml
└── src
    └── simple.cpp
```

```cpp
src/simple.cpp
#include "rclcpp/rclcpp.hpp"
int main(int argc, char * argv[]) {
  rclcpp::init(argc, argv);

  auto node = rclcpp::Node::make_shared("simple_node");

  rclcpp::spin(node);

  rclcpp::shutdown();

  return 0;
}
```

- #include "rclcpp/rclcpp.hpp" 指令允许使用 C++ 中的大部分 ROS2 类型和函数。
- rclcpp::init(argc, argv) 从启动此进程的参数中提取 ROS2 的一些命令行参数。
- 第 6 行创建了一个 ROS2 节点，其名称为 simple_node，类型为 std::shared_ptr。rclcpp::Node 类配备了许多别名和静态函数，以简化代码。SharedPtr 是 std::shared_ptr<rclcpp::Node> 的别名，make_shared 是 std::make_shared<rclcpp::Node> 的静态方法。

以下的几行代码是等价的，从一个纯 C++ 语法的语句到一个利用 ROS2 基础库的语句：

```
1. std::shared_ptr<rclcpp::Node> node = std::shared_ptr<rclcpp::Node>(
   new rclcpp::Node("simple_node"));
2. std::shared_ptr<rclcpp::Node> node = std::make_shared<rclcpp::Node>(
   "simple_node");
3. rclcpp::Node::SharedPtr node = std::make_shared<rclcpp::Node>(
   "simple_node");
4. auto node = std::make_shared<rclcpp::Node>("simple_node");
5. auto node = rclcpp::Node::make_shared("simple_node");
```

- 在这段代码中，spin 方法会阻塞程序，使其不会立即终止。在接下来的示例中将解释这个重要的功能。
- shutdown 方法用于在程序结束之前管理一个节点的关闭。

检查已准备好编译程序的 CMakeLists.txt 文件。为了清楚起见，已删除一些不相关的部分：

```
CMakeLists.txt
cmake_minimum_required(VERSION 3.5)
project(basics)

find_package(ament_cmake REQUIRED)
find_package(rclcpp REQUIRED)

set(dependencies
  rclcpp
)

add_executable(simple src/simple.cpp)
ament_target_dependencies(simple ${dependencies})

install(TARGETS
  simple
  ARCHIVE DESTINATION lib
  LIBRARY DESTINATION lib
  RUNTIME DESTINATION lib/${PROJECT_NAME}
)

if(BUILD_TESTING)
  find_package(ament_lint_auto REQUIRED)
  ament_lint_auto_find_test_dependencies()
endif()

ament_export_dependencies(${dependencies})
ament_package()
```

在这个文件中，可以识别出几个部分。

- 在文件的第一部分，使用 find_package 指定所需的包。ament_cmake 包是 colcon 命令所需要的，而 rclcpp 包是明确指定的。在创建一个 dependencies 依赖变量时，包含这个包依赖的包是一个好习惯，因为我们需要多次使用这个列表。
- 对于每一个可执行文件：

编译：使用 add_executable 进行编译，并指定结果的名称及其源代码。还要使用 ament_target_dependencies 使其他包的头文件和库对当前目标可用。由于没有额外的库的依赖，因此只使用 ament_target_dependencies 就可以了。

安装：使用 install 指明要安装生成程序的位置，一个单独的安装指令将适用于包的程序和库。通常情况下，会安装部署和运行程序所需的所有内容。如果没有安装，它就不存在。

编译工作空间：

```
cd ~/bookros2_ws

colcon build --symlink-install
```

由于我们已经创建了一个新程序，因此我们需要重新加载一个工作空间。打开一个新终端并执行：

```
$ ros2 run my_package simple
```

让我们看看会发生什么：完全没有任何反应。Simple 节点的计算图见图 2.4。

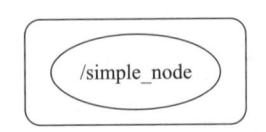

图 2.4　Simple 节点的计算图

在内部，我们的程序在 spin 语句中被阻塞，等待我们使用"Ctrl+C"结束程序。在执行此操作之前，请在另一个终端中检查该节点是否已经被创建。

```
$ ros2 node list
/simple_node
```

我们已经描述了如何从零开始创建一个包。从现在开始，我们将使用从上一章中的代码仓库下载的包。这样可以更高效地推进，不会因为构建包过程中的小错误而受阻，因为这些错误在这个阶段可能成为比较大的阻碍。

2.3　分析 br2_basics 软件包

一旦详细了解了这个过程，我们就继续分析 br2_basics 包的内容，它包含了更多有趣的节点。这个包的结构如下所示，完整的源代码可以在附录和本书的代码仓库中找到：

```
Package br2_basics

br2_basics
├── CMakeLists.txt
├── config
│   └── params.yaml
├── launch
│   ├── includer_launch.py
│   ├── param_node_v1_launch.py
│   ├── param_node_v2_launch.py
│   ├── pub_sub_v1_launch.py
│   └── pub_sub_v2_launch.py
├── package.xml
└── src
    ├── executors.cpp
    ├── logger_class.cpp
    ├── logger.cpp
    ├── param_reader.cpp
    ├── publisher_class.cpp
    ├── publisher.cpp
    └── subscriber_class.cpp
```

2.3.1 控制迭代执行

上一节描述了一个包含节点的程序，这个节点实际上没有做什么事情。程序 src/logger.cpp 会更有趣，因为它显示了更多的活动。

```
src/logger.cpp
auto node = rclcpp::Node::make_shared("logger_node");

rclcpp::Rate loop_rate_period(500ms);
int counter = 0;

while (rclcpp::ok()) {
  RCLCPP_INFO(node->get_logger(), "Hello %d", counter++);

  rclcpp::spin_some(node);
  loop_rate_period.sleep();
}
```

这段代码展示了以固定频率执行任务的一种常见的方法，这在执行某些控制的程序中经常见到。控制循环是在 while 循环中完成的，通过使用一个 rclcpp::Rate 对象来控制速率，使控制循环停止足够长的时间以适应所选的速率。

这段代码使用了 spin_some，而不是到目前为止我们使用的 spin。两者都用于管理到达节点的消息，调用应处理它们的函数。spin 会阻塞，因为需要等待新的消息，而 spin_some 则不会阻塞，因为没有消息立即返回。

代码的其他部分使用了 RCLCPP_INFO 打印信息的宏。它与 printf 非常相似，将节点的日志记录器（节点内部用于记录日志的对象，通过 get_logger 方法获得）作为第一个参数传递。这些消息会显示在屏幕上，也会发布到 /rosout 消息主题。

通过输入以下命令运行此程序：

```
$ ros2 run br2_basics logger
[INFO] [1643264508.056814169] [logger_node]: Hello 0
[INFO] [1643264508.556910295] [logger_node]: Hello 1
...
```

程序开始打印消息，内容包括消息的严重性级别、时间戳、产生它的节点和具体消息内容。

正如我们之前所说，RCLCPP_INFO 还向 /rosout 消息主题发布一个 rcl_interfaces/msg/Log 类型的消息，Logger 节点的计算图如图 2.5 所示。所有的节点都有一个消息发布器，将生成的输出信息发送到这个节点。当没有控制台来查看这些消息时，这个功能就非常有用了。

```
          /logger_node
              2Hz
```
→ /rosout
rcl_interfaces/msg/Log

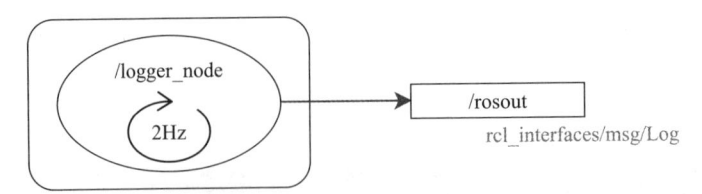

图 2.5　Logger 节点的计算图

借此机会了解如何查看发布到主题的消息：

```
$ ros2 topic echo /rosout
stamp:
    sec: 1643264511
    nanosec: 556908791
level: 20
name: logger_node
msg: Hello 7
file: /home/fmrico/ros/ros2/bookros2_ws/src/book_ros2/br2_basics/src/logger.cpp
function: main
line: 27
---
stamp:
    sec: 1643264512
    nanosec: 57037520
level: 20
...
```

查看 rcl_interfaces/msg/Log 类型消息的定义，以验证显示的字段是否为这种消息类型的字段。line 字段会告诉我们消息是从源代码中的哪一行产生的，以方便调试。

```
$ ros2 interface show rcl_interfaces/msg/Log
```

最后，使用 rqt_console 工具查看发布在 /rosout 的消息，如图 2.6 所示。当多个节点向 /rosout 消息主题发布消息时，这个工具很有用，如图 2.7 所示，可以按节点、严重性级别等进行筛选。

```
$ ros2 run rqt_console rqt_console
```

通过改变 loop_rate 对象创建的时间来测试不同的频率，并将其更改为 100ms 或 1s，使得控制循环分别以 10Hz 或 1Hz 运行。

每次修改后都不要忘记编译。使用 --packages-select 选项只编译修改过的包，从而节省时间。

```
         ┌─────────────┐
         │ /logger_node│
         │     ↓       │         ┌─────────┐
         │    2Hz      │────────▶│ /rosout │
         │             │         └─────────┘
         └─────────────┘              │
                                      │ rcl_interfaces/msg/Log
                                      ▼
                                ┌─────────────┐
                                │ rqt_console │
                                └─────────────┘
```

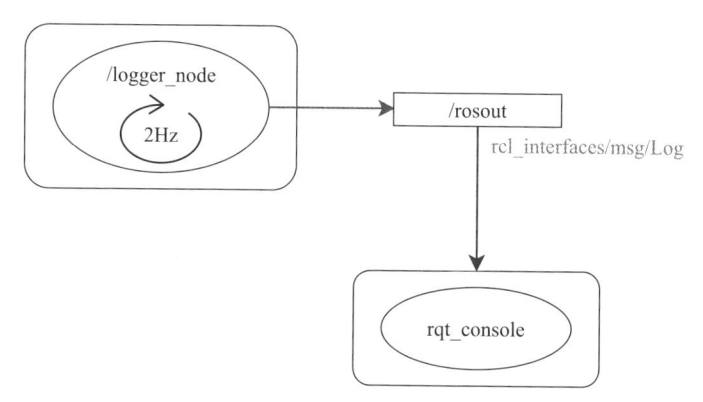

图 2.6　rqt_console 订阅 /rosout 消息主题，接收由 Logger 节点产生的消息

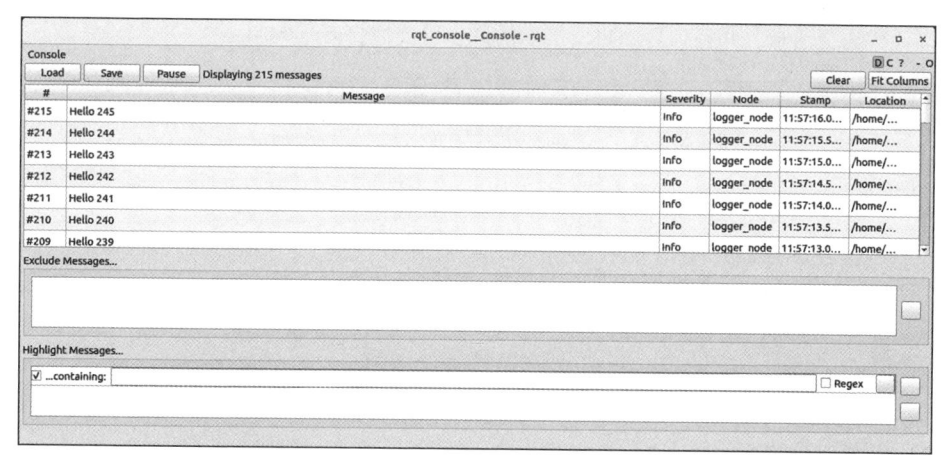

#	Message	Severity	Node	Stamp	Location
#215	Hello 245	Info	logger_node	11:57:16.0...	/home/...
#214	Hello 244	Info	logger_node	11:57:15.5...	/home/...
#213	Hello 243	Info	logger_node	11:57:15.0...	/home/...
#212	Hello 242	Info	logger_node	11:57:14.5...	/home/...
#211	Hello 241	Info	logger_node	11:57:14.0...	/home/...
#210	Hello 240	Info	logger_node	11:57:13.5...	/home/...
#209	Hello 239	Info	logger_node	11:57:13.0...	/home/...

图 2.7　rqt_console 工具

```
$ cd ~/bookros2_ws
$ colcon build --symlink-install --packages-select br2_basics
```

从这里开始将省略 cd 命令。必须从工作空间的根目录中执行所有编译。

深入了解：日志

ROS2 拥有一个日志系统，它允许按照不同的严重性级别生成日志消息：DEBUG、INFO、WARN、ERROR 或 FATAL。为此，请使用宏 RCLCPP_[LEVEL] 或 RCLCPP_[LEVEL]_STREAM 来使用文本流。

默认情况下，除了发送到 /rosout 消息主题外，严重性级别为 INFO 或更高

的消息将显示在标准输出上。你可以配置日志记录器以设定另一个最小的严重性级别,以在标准输出上显示:

```
$ ros2 run br2_basics logger --ros-args --log-level debug
```

当一个应用中有许多节点时,建议使用如 rqt_console 这样的工具,它允许选择节点和严重性。

延伸阅读:

- https://docs.ros.org/en/humble/Concepts/About-Logging.html

可以在 src/logger_class.cpp 程序中看到迭代执行任务的第二种策略。此外,我们展示了在 ROS2 中非常普遍的一种做法,即实现从 rclcpp::Node 继承的节点。这种方法可以让代码更清晰,并且为后面将要展示的许多可能性打开了大门:

```
src/logger_class.cpp
class LoggerNode : public rclcpp::Node
{
public:
  LoggerNode() : Node("logger_node")
  {
    counter_ = 0;
    timer_ = create_wall_timer(
      500ms, std::bind(&LoggerNode::timer_callback, this));
  }

  void timer_callback()
  {
    RCLCPP_INFO(get_logger(), "Hello %d", counter_++);
  }

private:
  rclcpp::TimerBase::SharedPtr timer_;
  int counter_;
};

int main(int argc, char * argv[]) {
  rclcpp::init(argc, argv);

  auto node = std::make_shared<LoggerNode>();

  rclcpp::spin(node);

  rclcpp::shutdown();
  return 0;
}
```

控制循环由一个计时器控制。这个计时器以所需的频率产生一个事件。当这个事件发生时,它调用处理该事件的回调。它的优点是在节点内部调整它应该执行的频率,而不把这个决定委托给外部代码。节点知道它们应该运行的频率。

要编译这些程序,CMakeLists.txt 中的相关行是:

- 对于每个可执行文件,有一个 add_executable 和它对应的 ament_target_dependencies。

- 一个带有所有可执行文件的 install 指令。

```
CMakeLists.txt
add_executable(logger src/logger.cpp)
ament_target_dependencies(logger ${dependencies})

add_executable(logger_class src/logger_class.cpp)
ament_target_dependencies(logger_class ${dependencies})
install(TARGETS
  logger
  logger_class
  ...
  ARCHIVE DESTINATION lib
  LIBRARY DESTINATION lib
  RUNTIME DESTINATION lib/${PROJECT_NAME}
)
```

```
$ ros2 run br2_basics logger_class
```

构建包并运行这个程序,你会看到效果与前一个程序相同。在创建计时器时,尝试通过在 create_wall_timer 中设置不同的时间来调整频率。

2.3.2 发布和订阅

对节点进行扩展,使其不再向屏幕打印消息,而是向名为 /counter 的消息主题发布连续数字,一个 Publisher 节点的计算图见图 2.8。通过 ros2 interface 命令的 list 和 show 选项,我们选择了 std_msgs/msg/Int32 作为这个任务的消息类型。

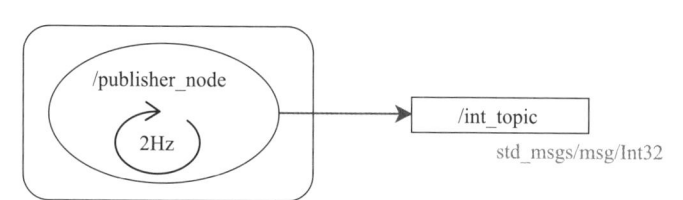

图 2.8 Publisher 节点的计算图

使用消息时,需要引入相应的头文件。对于 std_msgs/msg/Int32 这种消息,从其名称可以直观地得知要引入的头文件。只需要在大写字母前加一个空格并全部转为小写即可。这种命名方式也很直接。

```
// For std_msgs/msg/Int32
#include "std_msgs/msg/int32.hpp"

std_msgs::msg::Int32 msg_int32;


// For sensor_msgs/msg/LaserScan
#include "sensor_msgs/msg/laser_scan.hpp"

sensor_msgs::msg::LaserScan msg_laserscan;
```

现在我们来看一下 PublisherNode 的代码。

```
src/publisher_class.cpp
class PublisherNode : public rclcpp::Node
{
public:
  PublisherNode() : Node("publisher_node")
  {
    publisher_ = create_publisher<std_msgs::msg::Int32>("int_topic", 10);
    timer_ = create_wall_timer(
      500ms, std::bind(&PublisherNode::timer_callback, this));
  }

  void timer_callback()
  {
    message_.data += 1;
    publisher_->publish(message_);
  }
private:
  rclcpp::Publisher<std_msgs::msg::Int32>::SharedPtr publisher_;
  rclcpp::TimerBase::SharedPtr timer_;
  std_msgs::msg::Int32 message_;
};
```

让我们讨论以下几个关键点：
- 我们将使用 std_msgs/msg/Int32 消息。从这个名称我们得知：
  1）头文件是 std_msgs/msg/int32.hpp。
  2）数据类型是 std_msgs::msg::Int32。
- 创建一个消息发布者，它负责创建主题（如果不存在）和发布消息。通过这个对象，我们可以知道有多少订阅者在监听某个主题。我们使用的是 rclcpp::Node 的 create_publisher 公共方法，它返回一个指向 rclcpp::Publisher 的 shared_ptr 指针。它需要消息主题名称和 rclcpp::QoS 对象作为参数。这个 QoS 类的构造函数接收一个代表输出消息队列大小的整数，C++ 编译器会自动处理。稍后会看到，这里我们选择不同的 QoS（服务质量）。
- 我们生成了一个 std_msgs::msg::Int32 的消息，这个消息只有一个字段。每隔 500ms，我们在定时器的回调中递增这个字段，并使用发布器的 publish 方法发送消息。

深入了解：ROS2 的 QoS

QoS 在 ROS2 里是一个重要且有价值的特性，也是导致故障的一个可能点，因此必须很好地理解它。下面的资料详细解释了可以设置哪些 QoS 策略以及它们的含义。以下是如何在 C++ 中设置 QoS 策略的示例：

```
publisher = node->create_publisher<std_msgs::msg::String>(
  "chatter", rclcpp::QoS(100).transient_local().best_effort());
```

默认	可靠	易失	保留最新
服务	可靠	易失	正常队列
传感器	最大努力	易失	小队列
参数	可靠	易失	大队列

每个发布者都指定其 QoS 设置，每个订阅者也可以指定。但有些 QoS 设置之间是不兼容的，这样会导致订阅者收不到消息。

QoS 持久性配置的兼容性		订阅者	
		易失	瞬态本地
发布者	易失	易失	无连接
	瞬态本地	易失	瞬态本地

QoS 可靠性配置的兼容性		订阅者	
		最大努力	可靠
发布者	最大努力	最大努力	无连接
	可靠	最大努力	可靠

真正的标准应该是发布者的 QoS 策略要比订阅者更宽松。例如，一个传感器的驱动程序在发布数据时应选择可靠的 QoS 策略。订阅者可以选择通信是否可靠，或者选择最大努力尝试。在这种情况下，发布者的设置可能是这样的：

```
publisher_ = create_publisher<sensor_msgs::msg::LaserScan>(
  "scan", rclcpp::SensorDataQoS().reliable());
```

订阅者可以使用相同的 QoS，或去掉可靠的部分。

延伸阅读：

- https://design.ros2.org/articles/qos.html
- https://discourse.ros.org/t/about-qos-of-images/18744/16

运行程序：

```
$ ros2 run br2_basics publisher_class
```

看看我们在这个主题上发布了什么：

```
$ ros2 topic echo /int_topic
data: 16
---
data: 17
```

```
---
data: 18
...
```

我们应该看到数据字段值递增的 std_msgs/msg/Int32 消息。

现在我们要实现一个订阅此消息的节点：

```
src/subscriber_class.cpp
class SubscriberNode : public rclcpp::Node
{
public:
  SubscriberNode() : Node("subscriber_node")
  {
    subscriber_ = create_subscription<std_msgs::msg::Int32>("int_topic", 10,
      std::bind(&SubscriberNode::callback, this, _1));
  }

  void callback(const std_msgs::msg::Int32::SharedPtr msg)
  {
    RCLCPP_INFO(get_logger(), "Hello %d", msg->data);
  }
private:
  rclcpp::Subscription<std_msgs::msg::Int32>::SharedPtr subscriber_;
};
```

在这段代码里，我们创建了一个 rclcpp::Subscription 类型的对象，订阅了同样的消息主题，使用了同样类型的消息。在创建它时，我们指定了每当有消息发布到这个主题时，都会触发回调函数，这个函数会接收到一个 shared_ptr 类型的 msg 参数。

把这个程序添加到 CMakeLists.txt，构建后，在一个终端运行 publisher_class 消息发布程序，在另一个终端运行消息订阅程序。Publisher 和 Subscriber 节点的计算图如图 2.9 所示。我们可以看到屏幕上会显示接收到的主题消息。

```
$ ros2 run br2_basics subscriber_class
```

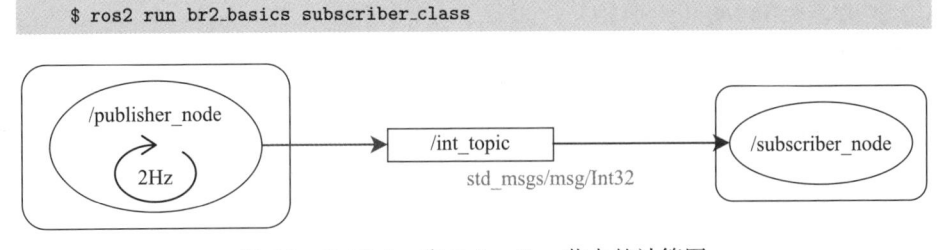

图 2.9 Publisher 和 Subscriber 节点的计算图

2.3.3 启动器

到现在为止，我们知道可以通过 ros2_run 来运行一个程序。在 ROS2 中，还有另一种方法是使用 ros2_launch 命令，并结合一个名为启动器的文件，该文件指定应该运行哪些程序。

启动器文件是用 Python[⊖]编写的，主要用来指定要执行的程序及其选项或参数。一个启动器还可以包含另一个启动器，支持复用已有的启动器。

需要启动器的原因是，一个机器人应用由多个节点组成，这些节点需要同时被启动。逐一启动并为每一个节点调整特定参数以使节点协作，这可能会很烦琐。

每个软件包的启动器都放在该包的 launch 目录下，文件名通常以 _launch.py 或 .launch.py 结尾。就像 ros2 run 能够自动识别软件包中的可执行文件一样，ros2 launch 也能做同样的事情，即识别出可用的启动器。

在技术实现角度来看，一个启动器是一个 Python 程序，其中包含一个 generate_launch_description() 函数，该函数返回一个 LaunchDescription 对象。一个 LaunchDescription 对象包含了一些动作，我们主要关注的动作如下。

- Node 动作：运行一个程序。
- IncludeLaunchDescription 动作：引入另一个启动器。
- DeclareLaunchArgument 动作：声明启动器参数。
- SetEnvironmentVariable 动作：设置一个环境变量。

看看我们如何同时启动发布者和订阅者节点。让我们分析 basics 包中的第一个启动器：

```
launch/pub_sub_v1_launch.py
from launch import LaunchDescription
from launch_ros.actions import Node

def generate_launch_description():
    pub_cmd = Node(
        package='br2_basics',
        executable='publisher',
        output='screen'
    )

    sub_cmd = Node(
        package='br2_basics',
        executable='subscriber_class',
        output='screen'
    )

    ld = LaunchDescription()
    ld.add_action(pub_cmd)
    ld.add_action(sub_cmd)

    return ld
```

与这个文件类似，在 launch/pub_sub_v2_launch.py 中有另一种实现方式，它的行为相同。你可以对比一下它们的不同之处。要使用启动器，我们必须安装启动器目录：

```
CMakeLists.txt
install(DIRECTORY launch DESTINATION share/${PROJECT_NAME})
```

⊖ 最新的 ROS2 发行版允许我们编写用 YAML 和 XML 编写的启动器。

构建工作空间并启动这个文件：

```
$ ros2 launch br2_basics pub_sub_v2_launch.py
```

在本节中，我们介绍了一些非常基础的启动器，其选项参数不多。随着深入学习，我们将看到更多复杂性的选项。

2.3.4 参数

节点使用参数来配置其操作。当你的程序需要配置信息时，就可以使用参数。这些参数可以是布尔型、整型、浮点型、字符串或者上述类型组成的数组。通常，在节点启动时，它会在运行中读取参数，并根据这些参数值来执行相应的操作。

设想一个节点，它的任务是利用粒子滤波器[9]来定位机器人。为了完成这个任务，它需要一些参数，比如粒子的最大数量或者从哪些消息主题中获取感测数据。这些配置信息不应该直接被嵌入到源代码中，因为一旦我们更换机器人或者更改环境条件，这些值可能就需要调整了。

看一个在启动时读取这些参数的节点。在 basics 包中创建一个 param_reader.cpp 文件：

```cpp
src/param_reader.cpp

class LocalizationNode : public rclcpp::Node
{
public:
  LocalizationNode() : Node("localization_node")
  {
    declare_parameter("number_particles", 200);
    declare_parameter("topics", std::vector<std::string>());
    declare_parameter("topic_types", std::vector<std::string>());

    get_parameter("number_particles", num_particles_);
    RCLCPP_INFO_STREAM(get_logger(), "Number of particles: " << num_particles_);

    get_parameter("topics", topics_);
    get_parameter("topic_types", topic_types_);

    if (topics_.size() != topic_types_.size()) {
      RCLCPP_ERROR(get_logger(), "Number of topics (%zu) != number of types (%zu)",
        topics_.size(), topic_types_.size());
    } else {
      RCLCPP_INFO_STREAM(get_logger(), "Number of topics: " << topics_.size());
      for (size_t i = 0; i < topics_.size(); i++) {
        RCLCPP_INFO_STREAM(
          get_logger(),
          "\t" << topics_[i] << "\t - " << topic_types_[i]);
      }
    }
  }

private:
  int num_particles_;
  std::vector<std::string> topics_;
  std::vector<std::string> topic_types_;
};
```

- 在节点中，必须使用 declare_parameter 这样的方法声明节点的所有参数，在声明中指定参数名称和默认值。
- 使用与 get_parameter 类似的函数获取其值，指定参数名称和参数值的存储位置。
- 有方法可以在代码块中一次性完成这些操作。
- 参数可以随时读取。你还可以订阅对这些参数的修改。但是，在启动时读取参数可以使代码更容易预测。

如果在不给参数赋值的情况下运行程序，就会发现默认值会生效：

```
$ ros2 run br2_basics param_reader
```

停止当前执行的程序，并执行我们的程序来给其中的某个参数赋值。参数设置以 --ros-args 开始，并使用 -p 选项来设置某个具体参数：

```
$ ros2 run br2_basics param_reader --ros-args -p number_particles:=300
```

再为剩余的两个字符串数组参数传入所需的值：

```
$ ros2 run br2_basics param_reader --ros-args -p number_particles:=300
-p topics:= '[scan, image]' -p topic_types:='[sensor_msgs/msg/LaserScan,
sensor_msgs/msg/Image]'
```

如果想在启动程序时设置参数值，可以这样做：

launch/param_node_v1.launch.py
```python
from launch import LaunchDescription
from launch_ros.actions import Node

def generate_launch_description():
    param_reader_cmd = Node(
        package='br2_basics',
        executable='param_reader',
        parameters=[{
            'particles': 300,
            'topics': ['scan', 'image'],
            'topic_types': ['sensor_msgs/msg/LaserScan', 'sensor_msgs/msg/Image']
        }],
        output='screen'
    )

    ld = LaunchDescription()
    ld.add_action(param_reader_cmd)

    return ld
```

虽然这种方法适合为少数参数赋值，但通常使用一个包含我们想要执行节点的参数值的文件更为方便。这就是在 ROS2 中采用的配置文件的方式。

我们选择了 YAML 作为配置文件的格式。这些文件一般都放置在包内的 config

目录下。与为 launch 目录所做的操作类似，在 CMakeLists.txt 中，必须标明这些配置文件以便安装。

CMakeLists.txt
```
install(DIRECTORY launch config DESTINATION share/${PROJECT_NAME})
```

我们需要探讨一个关键点：开发者为何不在自己的软件包中采用其他的目录结构？为何选择 config 目录而非 set up 或 startup，或是选择 launch 而不是其他选项？为何会选择将源文件放置于其他不同的文件结构中？为何选择 YAML 参数作为配置，而非纯文本或 XML，或是一个自定义的配置读取工具？为何要用启动器而非 bash 脚本？又为何不直接使用一个应用程序来一次性启动所有必要的节点？

显然，ROS2 的开发者完全有权做出不同的选择，但也确实存在一些广为人知的最佳实践。遵循这些实践的优势在于，当其他开发者想要使用你的代码时，他们可以更轻松地找到并确定关键部分。因此，我强烈推荐遵循这些惯例。这不仅使得你的代码能够为更多人所用，而且从长远来看，代码也更具可维护性，并且有可能吸引更多的合作伙伴。在商业环境中，这对于知识的传递和软件的维护都至关重要，尤其是当团队中的开发者不断变化时。

继续我们的例子。一个包含节点参数的文件可能看起来像这样：

config/params.yaml
```
localization_node:
  ros__parameters:
    number_particles: 300
    topics: [scan, image]
    topic_types: [sensor_msgs/msg/LaserScan, sensor_msgs/msg/Image]
```

执行程序，同时指定参数文件的位置。如果已经将 config 目录安装好并完成了编译，可以如下执行：

```
$ ros2 run br2_basics param_reader --ros-args --params-file
install/basics/share/basics/config/params.yaml
```

如果我们希望在一个启动器中读取该文件，我们将使用：

launch/param_node_v1.launch.py
```
def generate_launch_description():
    ...
    param_reader_cmd = Node(
        package='br2_basics',
        executable='param_reader',
```

launch/param_node_v1_launch.py
```
        parameters=[param_file],
        output='screen'
    )
```

2.3.5 执行器

由于 ROS2 中的节点是基于 C++ 的对象，这意味着一个进程可以有多个节点。实际上，在许多情况下，这样做是非常有利的，因为当通信在同一进程内时，可以通过使用共享内存策略来加速通信。这样做的另一个好处是，如果所有的节点都在同一个程序中，就可以简化节点的部署。但这样也会导致问题：一个节点的失败可能导致同一进程的所有节点都停止运行。

在 ROS2 中，你可以通过多种方式在同一个进程中运行多个节点。其中，最被推荐的方法是利用执行器。一个执行器是一个可以添加节点的对象，这样它们就可以一起被执行。以下是一个例子：

Single thread executor
```cpp
int main(int argc, char * argv[]) {
  rclcpp::init(argc, argv);

  auto node_pub = std::make_shared<PublisherNode>();
  auto node_sub = std::make_shared<SubscriberNode>();

  rclcpp::executors::SingleThreadedExecutor executor;

  executor.add_node(node_pub);
  executor.add_node(node_sub);

  executor.spin();

  rclcpp::shutdown();
  return 0;
}
```

Multi thread executor
```cpp
auto node_pub = std::make_shared<PublisherNode>();
auto node_sub = std::make_shared<SubscriberNode>();

rclcpp::executors::MultiThreadedExecutor executor(
    rclcpp::executor::ExecutorArgs(), 8);

executor.add_node(node_pub);
executor.add_node(node_sub);

executor.spin();
}
```

在这两段代码中，我们都创建了一个执行器，并向其添加了两个节点（运行在同一进程中的 Publisher 和 Subscriber 节点的计算图如图 2.10 所示），以便 spin() 调用能够同时处理这两个节点。这两者的主要区别在于，一个使用单一线程进行管理，而另一个使用八个线程优化处理器的性能。

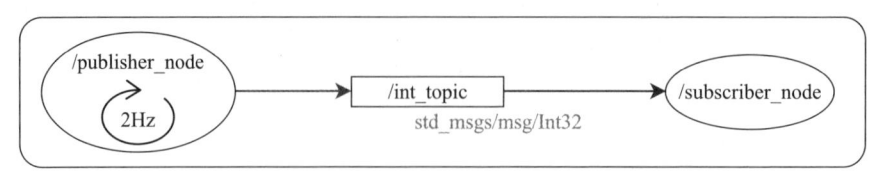

```
/publisher_node  ──→  /int_topic  ──→  /subscriber_node
      2Hz              std_msgs/msg/Int32
```

图 2.10　运行在同一进程中的 Publisher 和 Subscriber 节点的计算图

2.4　模拟机器人设置

到目前为止，我们已经介绍了基础包，该包为我们展示了 ROS2 的基础元素，包括如何创建节点、进行消息发布和消息订阅。值得注意的是，ROS2 并不仅仅是一个通信中间件，还是一个专门为机器人编程设计的中间件，本书的目的就是为机器人设计行为。因此，我们需要有一个机器人。然而，机器人的成本相对比较昂贵。例如，购买一个真实的、配备了激光和 RGBD 摄像头的 Kobuki 机器人（也称为 turtlebot 2）可能需要大约 1000 欧元。被视为专业级别的机器人的价格可以高达数万欧元。鉴于并非所有读者都计划购买一个实体机器人来运行 ROS2，我们将在模拟器中使用 Tiago 机器人（见图 2.11）。

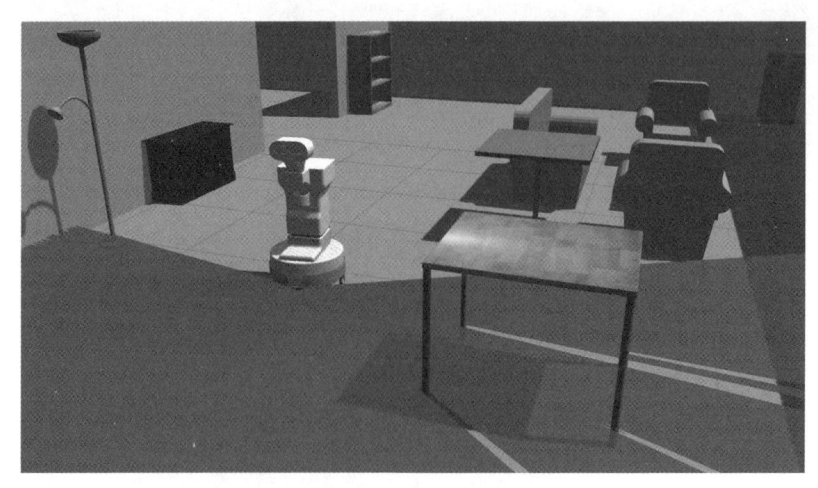

图 2.11　在模拟器 Gazebo 中进行模拟的 Tiago 机器人（见彩插）

PAL Robotics 公司的 Tiago 机器人（型号为"铁"）包含一个装有距离传感器的差动基座和一个带有手臂的躯干，而机器人的头部则配备了一个 RGBD 摄像头。

我们已经将所需的包添加到工作空间中，其中已经包括了在 Gazebo（ROS2 中的参考模拟器之一）中进行模拟 Tiago 机器人所需的包。所以只需要使用在 br2_tiago 包中创建的启动器即可：

```
$ ros2 launch br2_tiago sim.launch.py
```

我们提供了几种可用的世界（你可以查看 ThirdParty/pal_gazebo_worlds/worlds 文件夹下的内容）。默认加载的世界是 home.world。如果你想使用其他的，那么你可以使用启动器的 world 参数，如下面的例子所示：

```
$ ros2 launch br2_tiago sim.launch.py world:=factory
$ ros2 launch br2_tiago sim.launch.py world:=featured
$ ros2 launch br2_tiago sim.launch.py world:=pal_office
$ ros2 launch br2_tiago sim.launch.py world:=small_factory
$ ros2 launch br2_tiago sim.launch.py world:=small_office
$ ros2 launch br2_tiago sim.launch.py world:=willow_garage
```

当你第一次使用一个机器人并刚刚启动它的驱动程序或模拟器时，你可以做的第一件事是查看它作为发布者或订阅者提供的主题有哪些。这将是我们用来接收机器人的信息和发送命令的接口。打开一个新的终端并执行：

```
$ ros2 topic list
```

这将是与机器人的传感器和执行器交互的主要接口。图 2.14 展示了程序员可以与模拟机器人互动的节点和主题的概览。

- 实际上，几乎所有的节点都在 Gazebo 模拟器的进程内。在其外部只有两个节点：

    /twist_mux：创建若干个消息主题订阅者，接收来自不同源（移动设备、平板、键盘、导航等）的机器人速度的信息。

    /robot_state_publisher：ROS2 中的一个标准节点，它从 URDF 文件中读取机器人的描述，并订阅每个机器人关节的状态。除了在 URDF 中发布这个描述，它还在 TF 系统中创建和更新机器人的坐标系（我们将在下一章解释 TF 系统），这是一个表示和链接机器人中不同几何参考轴的系统。
- 左边的节点负责传感器。它们发布机器人的摄像头、惯性测量单元、激光和声纳的信息。其中最复杂的节点是摄像头节点，它是一个 RGBD 传感器，因为它分别发布深度和 RGB 图像。每个图像都有一个关联的名称为 camera_info 的消息主题，其中包含机器人摄像头的内参。对于每个传感器，都使用标准的消息类型来提供信息。
- 底部的节点使用相同的接口来移动头部和躯干。它们都使用 ros2_control 包中的 joint_trajectory_controller。
- 右侧的节点负责以下内容：

/joint_state_broadcaster：发布机器人每个关节的状态。

/mobile_base_controller：使机器人基座按接收到的速度指令运动。此外，它还发布基座的估算位移。

首先，遥控机器人进行移动。ROS2 有多个软件包可以从键盘、PS 或 XBox 控制器或手机接收指令，并将 geometry_msgs/msg/Twist 消息发布到一个消息主题中。在这种情况下，我们将使用 teleop_twist_keyboard。这个程序通过 stdin 标准输入接收按键操作，并发布 /cmd_vel 移动指令。

由于 teleop_twist_keyboard 的消息主题 /cmd_vel 与机器人的输入的任何消息主题都不匹配，因此我们必须做一个重映射。一个重映射（见图 2.12）允许在执行时（部署阶段）更改其主题的名称。在这种情况下，我们将执行 teleop_twist_keyboard，指示它不是在 /cmd_vel 中发布，而是发布到 /key_vel 主题中：

```
$ ros2 run teleop_twist_keyboard teleop_twist_keyboard --ros-args -r cmd_vel:=key_vel
```

```
/teleop_twist_keyboard ──▶ /cmd_vel ──重映射──▶ /key_vel ──▶ /twist_mux
                          geometry_msgs/           geometry_msgs/
                            msg/Twist                msg/Twist
```

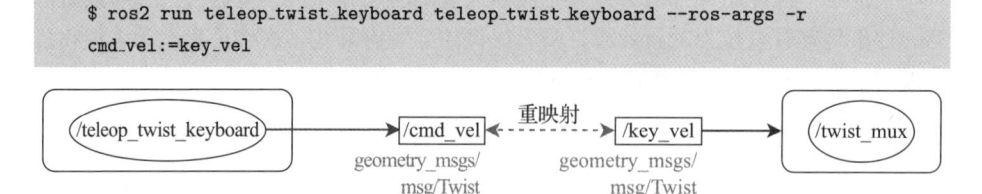

图 2.12　使用重映射连接 Tiago 和遥控器速度主题

现在，我们能够使用 teleop_twist_keyboard 所指示的按键来移动机器人了。

重映射消息主题是 ROS2 的一个重要特性，它允许来自其他开发者的不同 ROS2 程序协同工作。

现在我们可以开始观察机器人的传感器信息了。我们已经可以利用 ros2 topic 命令查看摄像头或激光的主题信息，这通过以下命令之一来实现：

```
$ ros2 topic echo /scan_raw
$ ros2 topic echo /head_front_camera/rgb/image_raw
```

但是显示传感器信息是很困难的，特别是当它如此复杂的时候。使用 --no-arr 选项，这样它就不会显示数据数组的内容。

```
$ ros2 topic echo --no-arr /scan_raw
$ ros2 topic echo --no-arr /head_front_camera/rgb/image_raw
```

仔细分析它显示的信息。在这两个消息中有一个共同的字段，这在带有感知信息的消息中是常见的,并且在许多类型的消息中都会重复,特别是那些以"Stamped"结尾的。它有一个类型为 std_msgs/msg/Header 的头部。正如我们刚刚看到的,消息可以通过组合基本类型(如 Int32、Float64、String)或已经存在的消息来定义。

头部信息在 ROS2 中对于处理传感器信息至关重要。当一个传感器驱动程序发布其数据的消息时,它使用头部信息来标记这个读数:

- 数据捕获的时间戳。即使一个消息收到或处理得晚,也可以将读数放在其相应的捕获时刻,同时支持某些延迟。
- 数据被采集的坐标系。坐标系是一个参考系,消息中包含的空间信息(如坐标、距离等)是有意义的。通常,每个传感器都有自己的坐标系(甚至有几个)。

一个机器人是通过一棵树来几何建模的,树的节点都代表机器人的一个坐标系。按照惯例,一个坐标系应该有一个单独的父坐标系和所有必要的子坐标系。父子坐标系之间的关系是通过一个包含平移和旋转的几何变换来实现的。坐标系通常出现在机器人的变化点上,比如连接机器人各部件的电机。

在 ROS2 中有一个叫作 TF 的系统,我们将在下一章解释,该系统通过 /tf 和 /tf_static 两个消息主题来维护这些关系:/tf 用于变化的几何变换;/tf_static 用于固定的变换。

ROS2 提供了多种工具以帮助我们展示传感器数据和几何信息,其中最受欢迎的可能是 RViz2。首先,在终端中输入以下命令来启动它:

```
$ ros2 run rviz2 rviz2
```

RViz2 是一款查看器,可用于展示消息主题中的信息。如果你初次打开 RViz2,那么它可能会显示为空。它只有一个网格,我们可以使用键盘和鼠标在其中进行导航。我们将逐步发现有关我们机器人的信息,RViz2 可视化展示 Tiago 机器人的 TF 树和传感器信息如图 2.13 所示。

- 在左边的 Displays 面板中,有一些 RViz2 的全局选项,我们必须在这里指定固定坐标系是什么,也就是右侧显示的 3D 可视化的坐标轴。现在,我们选择 base_footprint。按照 ROS2 的惯例,这个坐标系是机器人中心坐标系,位于地面上,是我们进行探索的一个好的起点。
- 在 Displays 面板中,我们将添加不同的可视化元素。首个元素是机器人的坐标系。单击"添加"(Add)按钮,找到"按显示类型"(By display type)选项卡,并选择 TF 元素即可。此时,所有的机器人坐标系都会立即显示。如果你觉得坐标系太多了,就可以在 Displays 面板中选中 TF 组件,取消勾选"所有启用"(All Enabled)选项,随后添加或移除所需的坐标系。

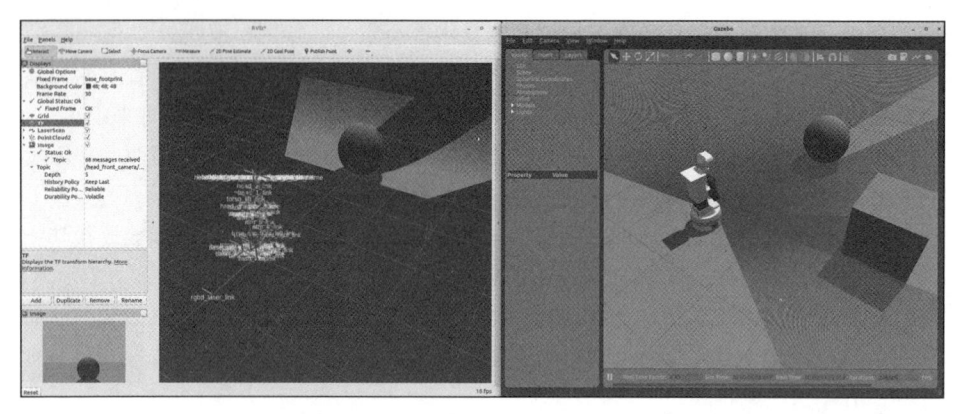

图 2.13 RViz2 可视化展示 Tiago 机器人的 TF 树和传感器信息（见彩插）

- 在 Gazebo 中添加几个元素。如果不这样做，我们也无法感知到太多信息。
- 添加机器人的激光信息。再次单击"添加"（Add）按钮，在"按消息主题"（By Topic）选项卡中，选择 /scan_raw 主题，该主题已经表明它负责处理激光扫描。在 Displays 面板上已添加的激光扫描元素中，我们可以查看信息并更改显示选项。此元素的选项如下：
  1）状态应该显示 ok，并且有一个计数器随着接收到的消息的增加而增加。如果显示出错，那么它通常会包含一些信息，这些信息可以帮助我们找出如何修复它的信息。
  2）消息主题与要显示的内容以及 RViz2 订阅该主题的 QoS 有关。如果我们没有看到任何内容，则可能是因为我们没有选择一个兼容的 QoS。
  3）从这里开始，其余的选项都是针对这种类型的消息的。我们可以改变代表激光器读数的点的大小、颜色，甚至可以改变使用的视觉元素。
- 像处理激光一样添加一个主题，该主题包含 PointCloud2（head_front_camera→depth_registered→points）的可视化信息，Tiago 机器人的计算图如图 2.14 所示。

使用遥控器移动机器人。在 RViz2 中，机器人的移动不会被感知，只有 odom 参考坐标系在移动。这是因为这个可视化的中心总是固定参考坐标系，我们现在把它设为 base_footprint。将固定参考坐标系更改为 odom，这是一个表示机器人启动时位置的参考坐标系。现在，我们可以看到机器人在其环境周围的移动情况。按照 ROS2 的惯例，odom→base_footprint 的变换收集了机器人驱动程序从起始点计算的平移和旋转。

通过这些，我们已经探索了模拟器机器人和管理机器人的各种工具的能力。

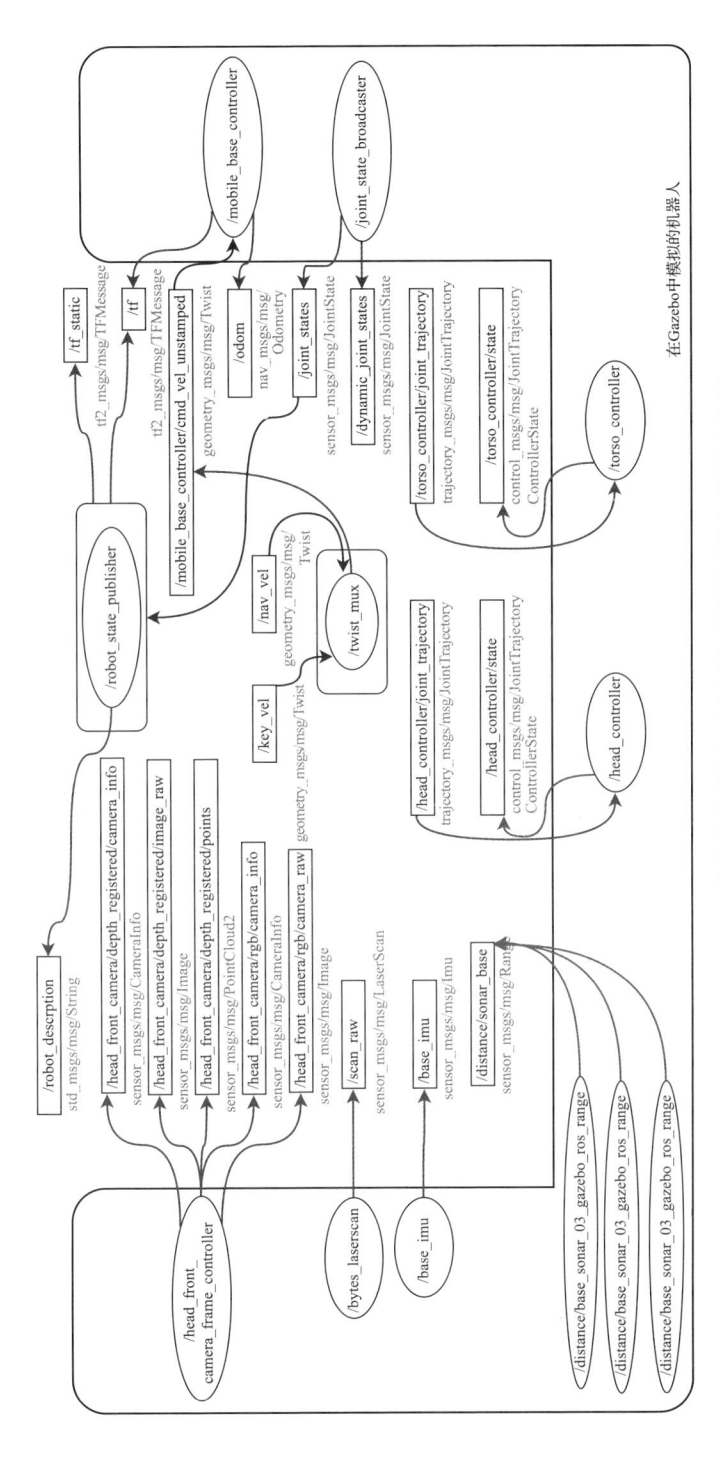

图 2.14 Tiago 机器人的计算图，图中展示了用到的相关消息主题

CHAPTER 3

第 3 章

第一个行为：用有限状态机避开障碍

本章旨在应用目前所展示的所有知识，以创造近似"智能"的行为。这个练习将整合我们之前学习的东西，并展示使用 ROS2 编程机器人是多么有效。此外，我们还将探讨机器人编程中的一些问题。

"碰撞与前进"（Bump and Go）的行为利用了机器人的传感器来检测其前方的障碍物。机器人向前移动，当它检测到障碍物时，它会后退并转动一段固定时间，然后再次向前移动。尽管这是一个简单的行为，但也推荐使用一种决策制定方法，因为我们的代码即使很简单，也可能在解决后续可能出现的问题时出现故障。在这种情况下，我们将使用有限状态机（Finite State Machine，FSM）。

有限状态机是一种数学计算模型，我们可以用它来定义机器人的行为。它由状态和转换组成。机器人在一个状态中不断产生输出，直到满足某个出站转换的条件，并转移到相应的目标状态。

应用有限状态机可以显著降低解决问题的复杂性，尤其是在实现简单行为时。试着想一想如何使用循环、if 语句、临时变量、计数器和计时器来解决"碰撞与前进"问题。这将是一个难以理解和跟踪其逻辑的复杂程序。一旦完成，此时添加一些额外条件可能会使你放弃我们已经完成的工作并重新开始。

而利用基于 FSM 的解决方案来解决"碰撞与前进"问题则很简单。考虑到机器人必须产生的不同输出（停止、前进、后退和转向），每个动作都将有自己的状态。现在考虑状态之间的转换（连接和条件），我们将得到一个类似于图 3.1 所示的 FSM 图。

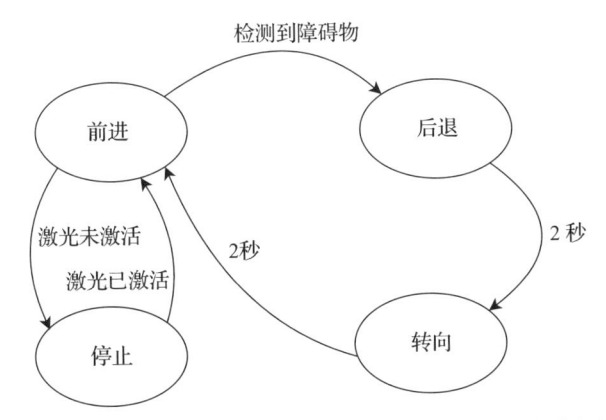

图 3.1 使用 FSM 解决"碰撞与前进"问题的有限状态机图

3.1 感知和执行模型

本节分析了解决"碰撞与前进"问题所使用的感知以及可以产生的动作。

在两种模型中，首先必须定义所使用的几何约定：

- ROS2 使用国际单位制（SI）进行度量。对于不同的尺寸，我们将考虑米（m）、秒（s）和弧度（rad）作为单位。线速度应为 m/s，旋转速度应为 rad/s，线性加速度应为 m/s^2 等等。
- 在 ROS2 中，我们采用右手坐标系[⊖]（见图 3.2 左半部分）：x 轴向前增长，y 轴向左增长，z 轴向上增长。如果我们将参考原点设在胸部位置，那么 x 值为负代表在我们后方，z 值为正代表在我们上方。
- 角度被定义为绕轴的旋转。围绕 x 轴的旋转被称为滚动，绕 y 轴的旋转被称为俯仰，绕 z 轴的旋转被称为偏航。
- 逆时针旋转时，角度增加（见图 3.2 右半部分）。角度 0 是向前的，π 是向后的，π/2 是向左的。

对于这个问题，我们将使用激光传感器的信息，我们在上一章中看到这个信息在 /scan_raw 消息主题中，其类型为 sensor_msgs/msg/LaserScan。通过输入以下命令来检查此消息格式：

```
$ ros2 interface show sensor_msgs/msg/LaserScan

# Single scan from a planar laser range-finder
#
# If you have another ranging device with different behavior (e.g. a sonar
```

⊖ https://www.ros.org/reps/rep-0103.html。

```
# array), please find or create a different message, since applications
# will make fairly laser-specific assumptions about this data

std_msgs/Header header   # timestamp in the header is the acquisition time of
                         # the first ray in the scan.
                         #
                         # in frame frame_id, angles are measured around
                         # the positive Z axis (counterclockwise, if Z is up)
                         # with zero angle being forward along the x axis

float32 angle_min        # start angle of the scan [rad]
float32 angle_max        # end angle of the scan [rad]
float32 angle_increment  # angular distance between measurements [rad]

float32 time_increment   # time between measurements [seconds] - if your scanner
                         # is moving, this will be used in interpolating pos
                         # of 3d points
float32 scan_time        # time between scans [seconds]

float32 range_min        # minimum range value [m]
float32 range_max        # maximum range value [m]

float32[] ranges         # range data [m]
                         # (Note: values < range_min or > range_max should be
                         # discarded)
float32[] intensities    # intensity data [device-specific units]. If your
                         # device does not provide intensities, please leave
                         # the array empty.
```

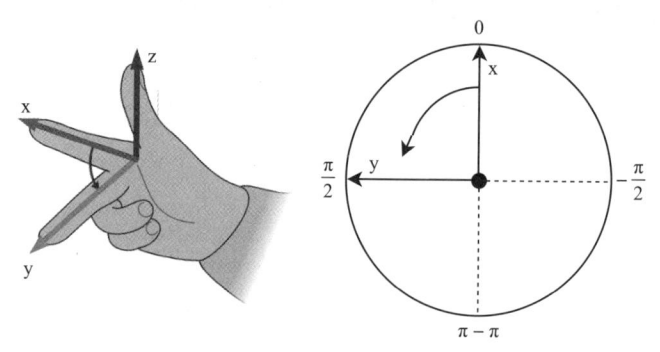

图 3.2　ROS 中的轴和角度约定（见彩插）

要查看这些激光消息中的一条（不显示读数内容），请启动模拟器并输入：

```
$ ros2 topic echo /scan_raw --no-arr
---
header:
  stamp:
```

```
      sec: 11071
    nanosec: 445000000
   frame_id: base_laser_link
angle_min: -1.9198600053787231
angle_max: 1.9198600053787231
angle_increment: 0.005774015095084906
time_increment: 0.0
scan_time: 0.0
range_min: 0.05000000074505806
range_max: 25.0
ranges: '<sequence type: float, length: 666>'
intensities: '<sequence type: float, length: 666>'
---
```

在图 3.3 中，我们可以看到这条消息的解释⊖。值得注意的是，在 ranges 字段中，存储着到障碍物的距离。std::vector 的第 0 个位置（在 C++ 中，消息中的数组表示为 std::vector）对应着角度 −1.9198，第 1 个位置则是这个角度加上增量，以此类推，直到该向量被完成。很容易验证，如果我们将角度范围（最大角度减去最小角度）除以增量，我们就得到了 666 个读数，即 ranges 向量的大小。

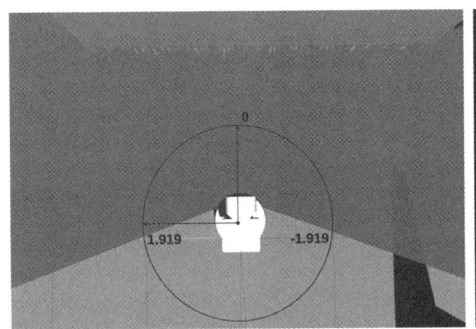

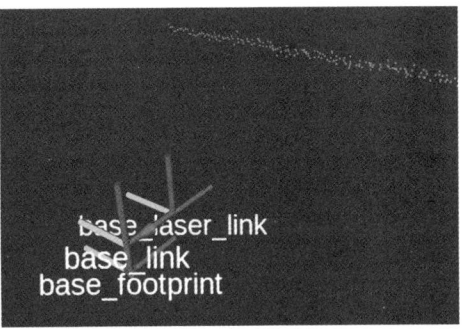

图 3.3 模拟的 Tiago 机器人中的激光扫描的解释（左图）；激光坐标系与其他主要坐标系的关系（右图）(见彩插)

大多数信息，特别是包含空间可解释的信息，都有一个包含时间戳和传感器坐标系的消息头。请注意，传感器可以安装在机器人的任何位置或方向上，甚至可以安装在某些移动部件上。传感器坐标系必须与其余部分的坐标系有几何上的连接

⊖ 基础激光链接坐标系（base_laser_link）是坐标系名称，它代表了机器人上激光雷达传感器的位置和方向。基础链接坐标系（base_link）是常用坐标系，它代表了机器人的中心，所有其他的机器人部件（例如传感器、执行器等）都相对于 base_link 坐标系进行定位。基础坐标系（base_footprint）是常用坐标系，它代表了机器人在二维平面（通常是地面）上的位置和方向。——译者注

（包括旋转和平移）。在许多情况下，我们需要将感知信息的坐标转换到同一坐标系中进行融合，通常是转换到基础坐标系（base_footprint），即机器人的中心、地面水平、指向前方。这些几何操作将在之后加以阐述。

在我们的问题中，我们只关心机器人前方（即角度为 0 的位置）是否存在障碍物，这恰好对应于 ranges 向量的中间位置的内容。我们可以使用传感器的原始坐标系，因为它与基础坐标系对齐，稍微前移和上移。

ROS 的一个重要特征是标准化。当社区就激光传感器所产生信息的编码格式达成共识时，所有激光驱动程序开发者都应该遵循此格式。这个共识意味着消息格式必须是通用的，以支持任何激光传感器。同样地，应用开发者必须利用这个消息来使他的程序正确运行，无论产生感官读数的传感器的特性如何。此方法的巨大优点在于我们可以使任何 ROS 程序与 ROS 支持的任何激光器一起工作，使软件具备可移植性。此外，经验丰富的 ROS 开发人员无须学习新的、由制造商定义的格式。最后，使用此格式可以为你提供各种过滤或监视激光信息的实用工具。这种方法适用于 ROS 中所有类型的传感器和执行器，这可能是 ROS 框架成功的原因之一。

关于此示例中的动作模型，我们将向 /nav_vel 消息主题发送机器人的平移和旋转速度数据，其类型为 geometry_msgs/msg/Twist。让我们来看一下此消息格式：

```
$ ros2 interface show geometry_msgs/msg/Twist
Vector3 linear
Vector3 angular

$ ros2 interface show geometry_msgs/msg/Vector3
float64 x
float64 y
float64 z
```

所有机器人都使用这种速度消息格式，从而允许在 ROS 中使用键盘、操纵杆、移动设备等实现通用的远程操作和导航功能，再次体现了 ROS 标准化。

geometry_msgs/msg/Twist 消息比我们机器人支持的消息更为通用。由于其为差速驱动机器人[一]，因此我们无法使其在 z 轴上移动（不能飞行）或仅使用两个轮子直接横向移动。如果我们有一架四旋翼飞行器，我们就可能进行更多的平移和旋转。我们只能前进或后退、旋转，或者将前进和旋转结合。因此，我们只能使用 linear.x 和 angular.z 的字段（绕 z 轴旋转，向左为正速度，如图 3.2 所示）。

㊀ 这种类型的机器人通常有两个驱动轮，它们以不同的速度旋转实现在平面上前进、后退和旋转的动作，而不能在垂直方向上移动。——译者注

3.2 计算图

"碰撞与前进"项目的计算图非常简单（见图 3.4）：一个节点订阅激光消息主题并向机器人发布速度指令。

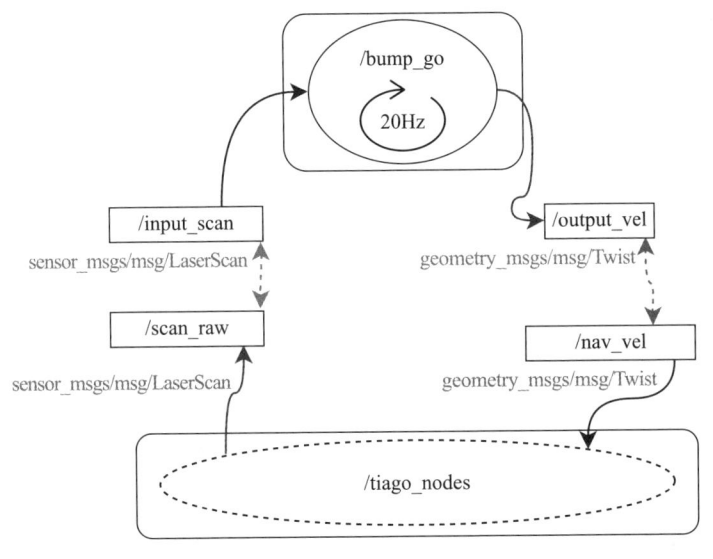

图 3.4 "碰撞与前进"项目的计算图

控制逻辑会解读输入的感知信息并生成控制命令。我们将使用有限状态机来实现该逻辑控制。该逻辑控制将以 20Hz 的迭代频率运行。执行频率取决于发布控制命令的频率，如果发布频率低于 20Hz，则会有一些机器人停止下来。这会非常方便，以保证实验室中不存在没被控制的机器人。通常，我们接收信息的频率与发布信息的频率不同。你必须处理这个问题。工程师不会抱怨有问题的情况——他们会修复这些问题。

如果我们希望软件能够在不同的机器人上运行，则无须为机器人指定特定的消息主题。在示例中，它订阅的消息主题是 /input_scan，并发布到 /output_vel。这些主题不存在或与我们模拟的机器人的主题不对应。在执行时（部署时），我们会重映射端口，将它们连接到特定机器人的实际的消息主题上。

让我们在这里进一步扩展一下。为什么我们要使用重映射（remap）而不是将主题的名称作为参数传递呢？这是许多 ROS2 开发人员提倡的一种替代方案。当一个节点并不总是具有相同的订阅者和发布者时，这种替代方法可能更方便，并且只能在一个 YAML 配置参数的文件中指定。

一个好的方法是，如果节点中发布者和订阅者的数量是已知的，则使用通用的主题名称,就像在这个示例中使用的那样,并执行重映射。使用常见的主题名称(例如 /cmd_vel 是许多机器人的常见主题)可能更好。经验丰富的 ROS2 程序员将在文档中阅读它使用的主题,使用 ros2 node info 找到它,然后快速使用重映射使其正常工作,而不是在配置文件中寻找正确的参数设置。

虽然本书主要使用 C++,但在本章中,我们将提供两种类似的实现,一个是 C++,另一个是 Python,每种实现都在不同的软件包中:br2_fsm_bumpgo_cpp 和 br2_fsm_bumpgo_py。两者都已经在之前创建的工作空间和本书的附录中。让我们从 C++ 实现开始。

3.3 "碰撞与前进"在 C++ 中的实现

br2_fsm_bumpgo_cpp 包具有以下结构:

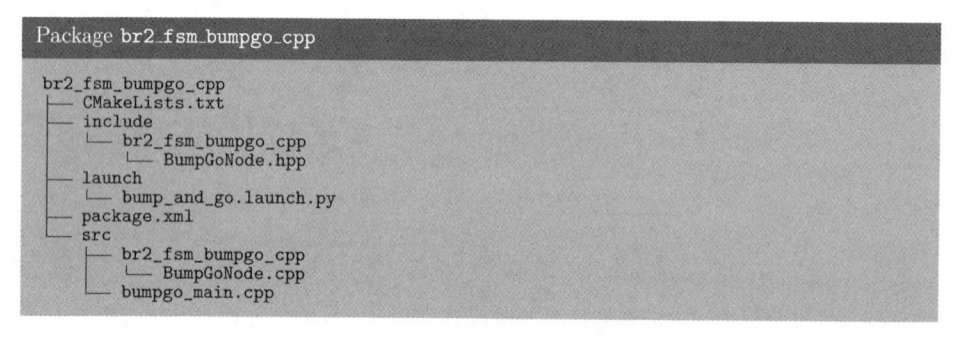

```
Package br2_fsm_bumpgo_cpp
br2_fsm_bumpgo_cpp
├── CMakeLists.txt
├── include
│   └── br2_fsm_bumpgo_cpp
│       └── BumpGoNode.hpp
├── launch
│   └── bump_and_go.launch.py
├── package.xml
└── src
    ├── br2_fsm_bumpgo_cpp
    │   └── BumpGoNode.cpp
    └── bumpgo_main.cpp
```

节点通常作为继承自 rclcpp::Node 的类实现,在一个与包名相匹配的命名空间中,将类声明和定义分开。在我们的例子中,类定义(BumpGoNode.cpp)位于 src/br2_fsm_bumpgo_cpp 目录下,其中头文件(BumpGoNode.hpp)位于 include/br2_fsm_bumpgo_cpp 目录下。这样,我们将程序的实现与节点的实现分开,这可以让我们使用多个不同的程序来实现节点。主程序的功能是在 src/bumpgo_main.cpp 中实例化节点并调用 spin() 函数。我们还提供了一个启动器(launch/bump_and_go.launch.py)来简化其执行过程。

在本书中,我们将分析软件包的部分代码,着重讲解不同的具体方面,以教授有趣的概念。这里不详尽展示所有代码,因为读者可以在自己的工作空间、代码仓库和附录中找到它。

3.3.1 执行控制

节点执行模型以 20Hz 的频率调用 control_cycle 方法。为此,我们在构造函数中声明一个计时器并启动它，以便每 50ms 调用一次 control_cycle 方法。控制逻辑使用有限状态机实现，将发布速度指令。

```
include/bump_go_cpp/BumpGoNode.hpp

class BumpGoNode : public rclcpp::Node
{
...
private:
  void scan_callback(sensor_msgs::msg::LaserScan::UniquePtr msg);
  void control_cycle();

  rclcpp::Publisher<geometry_msgs::msg::Twist>::SharedPtr vel_pub_;
  rclcpp::Subscription<sensor_msgs::msg::LaserScan>::SharedPtr scan_sub_;
  rclcpp::TimerBase::SharedPtr timer_;

  sensor_msgs::msg::LaserScan::UniquePtr last_scan_;
};
```

请注意激光回调头文件的详细信息。我们使用了 UniquePtr（std::unique_ptr 的别名），而不是 SharedPtr。ROS2 中的回调可以根据需要具有不同的签名。这些是回调的不同选择：

```
1. void scan_callback(const sensor_msgs::msg::LaserScan & msg);
2. void scan_callback(sensor_msgs::msg::LaserScan::UniquePtr msg);
3. void scan_callback(sensor_msgs::msg::LaserScan::SharedConstPtr msg);
4. void scan_callback(const sensor_msgs::msg::LaserScan::SharedConstPtr & msg);
5. void scan_callback(sensor_msgs::msg::LaserScan::SharedPtr msg);
```

其他一些签名可以让我们获得关于消息的信息（例如源和目的地的时间戳，以及发送者的唯一标识符），甚至是序列化的消息本身，但这仅用于一些特定情况。

到目前为止，我们使用了签名 1，但现在我们使用签名 2。请查看 scan_callback 中激光回调的实现。我们将获取此消息的属性，而不是复制消息（对于大消息可能会导致计算成本高）或共享指针，并将对数据的引用存储在 last_scan_ 中。这样，rclcpp 队列将不再需要管理它们的生命周期，从而节省时间。我们建议尽可能使用 UniquePtr 来提高节点的性能。

```
src/bump_go_cpp/BumpGoNode.cpp

BumpGoNode::BumpGoNode()
: Node("bump_go")
{
  scan_sub_ = create_subscription<sensor_msgs::msg::LaserScan>(
    "input_scan", rclcpp::SensorDataQoS(),
    std::bind(&BumpGoNode::scan_callback, this, _1));

  vel_pub_ = create_publisher<geometry_msgs::msg::Twist>("output_vel", 10);
  timer_ = create_wall_timer(50ms, std::bind(&BumpGoNode::control_cycle, this));
}

void
BumpGoNode::scan_callback(sensor_msgs::msg::LaserScan::UniquePtr msg)
{
  last_scan_ = std::move(msg);
}
```

```
src/bump_go_cpp/BumpGoNode.cpp
}

void
BumpGoNode::control_cycle()
{
  // Do nothing until the first sensor read
  if (last_scan_ == nullptr)
    return;

  vel_pub_->publish(...);
}
```

关于这个构造函数的另一个值得注意的细节是，消息发布使用默认的 QoS，即可靠和易失性。在订阅的情况下，我们将使用 rclcpp::SensorDataQoS()（一个使用最大努力、易失和适当的队列大小的服务质量的定义）。

通常而言，为了使消息通信兼容大部分场景，发布者的服务质量应该保证可靠，而订阅者的服务质量可以放宽标准至最大努力重试的机制。当创建传感器驱动时，使用 rclcpp::SensorDataQoS() 发布其读数不是一个好主意，因为如果订阅者需要可靠的 QoS 而发布者采用最大重试次数的话，通信将会失败。

最后，在 control_cycle 中要做的第一件事就是检查 last_scan_ 变量是否有效。这个方法可能会在第一个激光扫描的消息到达之前就被调用了。在这种情况下，本次迭代会被跳过执行。

3.3.2 实现有限状态机

在 C++ 类中实现有限状态机（FSM）并不复杂，只需要有一个 state_ 成员变量存储当前状态，我们可以将其编码为一个常量或枚举。此外，还需要有一个 state_ts_ 变量指示何时转换到当前状态，并允许使用超时进行状态之间的转换。

```
include/bump_go_cpp/BumpGoNode.hpp
class BumpGoNode : public rclcpp::Node
{
...
private:
  void control_cycle();

  static const int FORWARD = 0;
  static const int BACK = 1;
  static const int TURN = 2;
  static const int STOP = 3;
  int state_;
  rclcpp::Time state_ts_;
};
```

控制逻辑在 control_cycle 方法中执行，其执行频率为 20Hz。此方法不得包含任何死循环或长时间的暂停。

控制逻辑通常使用 switch 语句实现，每个 case 分支都有一个状态值。在下面的代码中，我们只展示了 FORWARD 状态值的处理情况。在这种情况下也有一个结构：首先是计算当前状态的输出（设置要发布的速率），然后检查每个转换条件。如果任何一个条件为真（条件满足），则将状态设置为新状态，并更新 state_ts_。

在声明一个消息类型变量时，所有字段默认设置为它们的默认值，或根据其类型设置为 0 或为空。这就是在完整代码中只分配非 0 字段的原因。

src/bump_go_cpp/BumpGoNode.cpp

```cpp
BumpGoNode::BumpGoNode()
: Node("bump_go"),
  state_(FORWARD)
{
  ...
  state_ts_ = now();
}

void
BumpGoNode::control_cycle()
{
  switch (state_) {
    case FORWARD:

      // Do whatever you should do in this state.
      // In this case, set the output speed.

      // Checking the condition to go to another state in the next iteration
      if (check_forward_2_stop())
        go_state(STOP);
      if (check_forward_2_back())
        go_state(BACK);

      break;
      ...
  }
}
void
BumpGoNode::go_state(int new_state)
{
  state_ = new_state;
  state_ts_ = now();
}
```

从实现的角度来看，这三个方法的代码很有趣。第一个是从前进（forward）状态转换为后退（back）状态的代码，它检查机器人前面是否有障碍物。正如我们之前所说，这是通过访问包含激光器测距读数向量的中心元素来完成的：

src/bump_go_cpp/BumpGoNode.cpp

```cpp
bool
BumpGoNode::check_forward_2_back()
{
  // going forward when detecting an obstacle
  // at 0.5 meters with the front laser read
  size_t pos = last_scan_->ranges.size() / 2;
  return last_scan_->ranges[pos] < OBSTACLE_DISTANCE;
}
```

第二个有趣的代码片段是当最后一个激光器读数被认为太旧时，从前进（forward）

状态转换到停止（stop）状态。rclcpp::Node 的 now 方法返回当前时间作为一个 rclcpp::Time。从最后一次读数的头部的时间开始，我们可以创建另一个 rclcpp::Time。它们之间的差是 rclcpp::Duration。为了进行比较，我们可以使用 seconds 方法，它以 double 形式返回秒数，或者可以像我们所做的那样，直接将其与另一个 rclcpp::Duration 进行比较。

src/bump_go_cpp/BumpGoNode.cpp

```cpp
bool
BumpGoNode::check_forward_2_stop()
{
  // Stop if no sensor readings for 1 second
  auto elapsed = now() - rclcpp::Time(last_scan_->header.stamp);
  return elapsed > SCAN_TIMEOUT;   // SCAN_TIMEOUT is set to 1.0
}
```

最后一个片段与前一个类似，但现在我们利用了更新后的 state_ts_ 变量，可以在 2s 后从后退（back）状态转换到转向（turn）状态。

src/bump_go_cpp/BumpGoNode.cpp

```cpp
bool
BumpGoNode::check_back_2_turn()
{
  // Going back for 2 seconds
  return (now() - state_ts_) > BACKING_TIME;
}
```

3.3.3 运行代码

到目前为止，我们仅局限于实现 BumpGoNode 节点的类。现在，我们必须了解在哪里创建这个类的对象来执行它。我们在主程序中创建节点的执行操作，并将其传递给一个 rclcpp::spin 阻塞调用，它将管理消息和计时器事件并调用它们的回调函数。

src/bumpgo_main.cpp

```cpp
int main(int argc, char * argv[])
{
  rclcpp::init(argc, argv);

  auto bumpgo_node = std::make_shared<br2_fsm_bumpgo_cpp::BumpGoNode>();
  rclcpp::spin(bumpgo_node);

  rclcpp::shutdown();

  return 0;
}
```

现在运行这个程序。打开一个终端来运行模拟器：

```
$ ros2 launch br2_tiago sim.launch.py
```

然后，开启另一个终端并运行程序，请考虑在命令行中指定参数：

- 将 input_scan 重映射为 /scan_raw，并将 output_vel 重新映射为 /nav_vel（使用 -r 选项）。
- 在使用模拟器时，将 use_sim_time 参数设置为 true。此设置会强制系统从模拟器发布的 /clock 消息主题中获取时间，而不是使用系统时钟。

```
$ ros2 run br2_fsm_bumpgo_cpp bumpgo --ros-args -r output_vel:=/nav_vel -r
input_scan:=/scan_raw -p use_sim_time:=true
```

观察机器人如何向前移动，直到检测到一个障碍物，然后进行避让操作。

由于在命令行中输入如此多的重映射参数很烦琐，因此我们创建了一个启动器，它用于指定必要的参数并对节点进行重映射。

```
launch/bump_and_go.launch.py
bumpgo_cmd = Node(package='br2_fsm_bumpgo_cpp',
    executable='bumpgo',
    output='screen',
    parameters=[{
      'use_sim_time': True
    }],
    remappings=[
      ('input_scan', '/scan_raw'),
      ('output_vel', '/nav_vel')
    ])
```

使用此启动器替代上一次的 ros2 run 指令，只需要输入以下内容：

```
$ ros2 launch br2_fsm_bumpgo_cpp bump_and_go.launch.py
```

3.4 "碰撞与前进"在 Python 中的实现

除了 C++ 之外，Python 也是 ROS2 官方支持的语言之一，通过 rclpy 客户端库被支持。本节将重现前一节所做的工作，但使用的是 Python。通过比较来验证两种语言在开发过程中的异同。此外，在之前的内容中已经解释了 ROS2 的原理，读者将能够在 Python 代码中识别出 ROS2 的元素，因为原理是相同的。

尽管我们提供了完整的包，但如果想从头开始创建一个包，我们仍然可以使用 ros2 pkg 命令来搭建一个框架。

```
$ ros2 pkg create --build-type ament_python br2_fsm_bumpgo_py --dependencies
sensor_msgs geometry_msgs
```

由于这是一个 ROS2 包，因此仍然有一个类似于 C++ 版本的 package.xml，但不再有 CMakeLists.txt，而是有一个 setup.cfg 和 setup.py，这是使用 distutils[⊖]的 Python

---

⊖ https://docs.python.org/3/library/distutils.html。

包的典型方式。

在这个包的根目录下,有一个同名的目录(它只有一个名为 __init__.py 的文件),表示该目录是一个 Python 模板。在此目录中创建 bump_go_main.py 文件。虽然在 C++ 中,将源代码分成几个文件是常见的,也是方便的,但 Python 实现倾向于将所有相关代码整合在同一个文件中。

3.4.1 执行控制

与之前的示例类似,我们首先展示忽略行为细节的代码,仅关注与处理 ROS2 概念相关的部分:

```
bump_go_py/bump_go_main.py
import rclpy

from rclpy.duration import Duration
from rclpy.node import Node
from rclpy.qos import qos_profile_sensor_data
from rclpy.time import Time

from geometry_msgs.msg import Twist
from sensor_msgs.msg import LaserScan


class BumpGoNode(Node):
    def __init__(self):
        super().__init__('bump_go')
        ...

        self.last_scan = None
        self.scan_sub = self.create_subscription(
            LaserScan,
            'input_scan',
            self.scan_callback,
            qos_profile_sensor_data)

        self.vel_pub = self.create_publisher(Twist, 'output_vel', 10)
        self.timer = self.create_timer(0.05, self.control_cycle)

    def scan_callback(self, msg):
        self.last_scan = msg

    def control_cycle(self):
        if self.last_scan is None:
            return

        out_vel = Twist()

        # FSM

        self.vel_pub.publish(out_vel)

def main(args=None):
    rclpy.init(args=args)

    bump_go_node = BumpGoNode()

    rclpy.spin(bump_go_node)

    bump_go_node.destroy_node()
    rclpy.shutdown()

if __name__ == '__main__':
    main()
```

前文提到，目标是创建一个节点，该节点订阅激光器读数并发出速度指令。控制周期以 20Hz 的频率执行，并根据最后接收到的读数来计算机器人的控制行为。因此，我们的代码将有一个订阅者、一个发布者和一个计时器。

这段代码与 C++ 代码类似：定义一个继承自 Node 的类，在主函数中实例化并调用 spin 函数。让我们来看一下细节：

- 从 Node 类继承后，我们调用基类构造函数来分配节点名称。Node 类和所有相关的数据类型（Time、Duration、QoS 等）都在 rclpy 包里，在启动时导入这些所需项。
- 消息类型也被导入，如初始化部分所示。
- 我们在构造函数中创建了发布者、订阅者和计时器。请注意，API 与 C++ 类似。此外，在 Python 中，我们可以访问预定义的服务质量（qos_profile_sensor_data）。
- 在激光消息的回调函数中，我们将最后接收到的消息存储在 self.last_scan 变量中，该变量在构造函数中被初始化为 None。通过这种方式，在 control_cycle 中验证我们没有接收到激光器读数。

3.4.2 实现有限状态机

将前一节中的 FSM 实现从 C++ 语言直接转换成 Python 时，核心逻辑并无显著差异。唯一的区别在于，为了获得当前时间，我们需要首先通过 get_clock 方法查询时钟：

```python
bump_go_py/bump_go_main.py
class BumpGoNode(Node):
    def __init__(self):
        super().__init__('bump_go')

        self.FORWARD = 0
        self.BACK = 1
        self.TURN = 2
        self.STOP = 3
        self.state = self.FORWARD
        self.state_ts = self.get_clock().now()

    def control_cycle(self):

        if self.state == self.FORWARD:
          out_vel.linear.x = self.SPEED_LINEAR

          if self.check_forward_2_stop():
            self.go_state(self.STOP)
          if self.check_forward_2_back():
            self.go_state(self.BACK)

        self.vel_pub.publish(out_vel)

    def go_state(self, new_state):
        self.state = new_state
        self.state_ts = self.get_clock().now()
```

或许，这个代码中最值得注意的一个方面是对时间和持续时间的处理方式，这与其 C++ 版本类似：

```
bump_go_py/bump_go_main.py

    def check_forward_2_back(self):
        pos = round(len(self.last_scan.ranges) / 2)
        return self.last_scan.ranges[pos] < self.OBSTACLE_DISTANCE

    def check_forward_2_stop(self):
        elapsed = self.get_clock().now() - Time.from_msg(self.last_scan.header.stamp)
        return elapsed > Duration(seconds=self.SCAN_TIMEOUT)

    def check_back_2_turn(self):
        elapsed = self.get_clock().now() - self.state_ts
        return elapsed > Duration(seconds=self.BACKING_TIME)
```

- Time.from_msg 函数允许从消息的时间戳创建一个 Time 对象。
- 使用节点的 get_clock().now() 方法获取当前时间。
- 时间之间的操作结果是一个 Duration 类型的对象，可以与另一个 Duration 类型的对象进行比较，例如 Duration(seconds=self.BACKING_TIME)，表示 2s 的持续时间。

3.4.3 运行代码

让我们看看如何在工作空间中构建和安装代码。首先，修改 setup.py 来适配我们的新程序：

```
setup.py

import os
from glob import glob

from setuptools import setup

package_name = 'br2_fsm_bumpgo_py'

setup(
    name=package_name,
    version='0.0.0',
    packages=[package_name],
    data_files=[
        ('share/ament_index/resource_index/packages',
            ['resource/' + package_name]),
        ('share/' + package_name, ['package.xml']),
        (os.path.join('share', package_name, 'launch'), glob('launch/*.launch.py'))
    ],
    install_requires=['setuptools'],
    zip_safe=True,
    maintainer='fmrico',
    maintainer_email='fmrico@gmail.com',
    description='BumpGo in Python package',
    license='Apache 2.0',
    tests_require=['pytest'],
    entry_points={
        'console_scripts': [
            'bump_go_main = br2_fsm_bumpgo_py.bump_go_main:main'
        ],
    },
)
```

现在重要的部分是 entry_points 参数。如上面的代码所示，添加之前显示的新程序。这样，我们就可以构建我们的包了。

```
$ colcon build --symlink-install
```

为了运行程序，请首先在终端中输入以下命令启动模拟器：

```
$ ros2 launch br2_tiago sim.launch.py
```

打开另一个终端并运行程序：

```
$ ros2 run br2_fsm_bumpgo_py bump_go_main --ros-args -r output_vel:=/nav_vel -r input_scan:=/scan_raw -p use_sim_time:=true
```

我们还可以使用类似 C++ 版本中的启动器，只需要输入：

```
$ ros2 launch br2_fsm_bumpgo_py bump_and_go.launch.py
```

建议的练习

1）修改"碰撞与前进"（Bump and Go）项目，使机器人能够感知前方、左右对角线上的障碍物。它不是总是向同一侧转，而是向没有障碍物的一侧转。

2）修改"碰撞与前进"（Bump and Go）项目，使机器人转向没有障碍物的角度或离其更远的障碍物。尝试两种方法：

- **开环控制**：在转向之前计算转向时间和速度。
- **闭环控制**：一直转到前方有足够的空间为止。

CHAPTER 4

第 4 章

TF 子系统

ROS 中的几何变换子系统 TF（简称 TF）堪称一大隐藏宝藏。该子系统允许定义不同的参考轴（也称为坐标系）以及它们之间的几何关系，即使这种关系在不断变化。一个坐标系中的任一坐标均可以重新计算到另一个坐标系中，而无须进行烦琐的手动计算。

在我教授 ROS 课程的经历中，那些曾经在没有 TF 的情况下处理过类似计算的学生，在学习了 TF 之后总是感到非常高兴。

TF 在 ROS 中的重要性源于需要对机器人的各个部件和组件进行几何建模。它在导航和定位以及操纵方面具有许多应用。它们已被用来定位一个建筑物中的几个摄像头或运动捕捉系统[一]。

机器人通过放在某个地方的传感器感知环境，并执行需要指定一些空间位置的动作。例如：
- 距离传感器（激光或 RGBD）生成一组点 (x, y, z) 以表示检测到的障碍物。
- 机器人通过指定目标位置 $(x, y, z, roll, pitch, yaw)$ 来移动其末端执行器。
- 机器人移动到地图上的一个点 (x, y, yaw)。

所有这些坐标都是相对于某个坐标系的参考。在机器人中，有多个坐标系（用于传感器、执行器等）。为了进行推理，必须了解这些坐标系之间的关系。例如，需要知道激光检测到的障碍物在机械臂参考轴上的坐标，以便避开它。坐标系之间的关系是一个坐标系相对于另一个坐标系的位移和旋转。在代数上，这是使用齐次坐标来表

一 https://github.com/MOCAP4ROS2-Project。

示坐标,并使用 RT 变换矩阵来表示坐标系之间的关系。基于已知点 P 在坐标系 A 中的坐标,即 P_A,我们可以使用变换矩阵 $RT_{A \to B}$,求出点 P 在坐标系 B 中的坐标 P_B。

$$P_B = RT_{A \to B} * P_A \qquad (4\text{-}1)$$

$$\begin{pmatrix} x_B \\ y_B \\ z_B \\ 1 \end{pmatrix} = \begin{pmatrix} R_{A \to B}^{xx} & R_{A \to B}^{xy} & R_{A \to B}^{xz} & T_{A \to B}^{x} \\ R_{A \to B}^{yx} & R_{A \to B}^{yy} & R_{A \to B}^{yz} & T_{A \to B}^{y} \\ R_{A \to B}^{zx} & R_{A \to B}^{zy} & R_{A \to B}^{zz} & T_{A \to B}^{z} \\ 0 & 0 & 0 & 1 \end{pmatrix} * \begin{pmatrix} x_A \\ y_A \\ z_A \\ 1 \end{pmatrix} \qquad (4\text{-}2)$$

除了这些操作的复杂性,在关节式机器人中这些关系的动态变化也值得关注。如果在高速动态变化的情况下,使用 t+0.01 秒时的变换来转换传感器在 t 时刻感知到的点,那么将会产生误差。

ROS2 实现了 TF 变换系统(现在称为 TF2,即第二个版本),它使用两个主题来接收变换,其消息类型为 tf2_msgs/msg/TFMessage。

```
$ ros2 interface show tf2_msgs/msg/TFMessage
geometry_msgs/TransformStamped[] transforms
    std_msgs/Header header
    string child_frame_id
    Transform transform
        Vector3 translation
            float64 x
            float64 y
            float64 z
        Quaternion rotation
            float64 x 0
            float64 y 0
            float64 z 0
            float64 w 1
```

- /tf 消息主题用于动态变化的变换,比如机器人的关节变换就在这里指定。默认情况下,它们的有效时间很短(10s)。例如,由驱动关节连接的坐标系关系就在此处发布。
- /tf_static 消息主题用于不随时间变化的变换。这个主题有一个 QoS 设置为瞬态本地(transient_local),所以任何订阅这个主题的节点都会接收到迄今为止发布的所有变换。通常,发布到这个主题的变换不会随时间变化,比如机器人的几何结构。

机器人的坐标系以 TF 树的形式组织,其中每个 TF 最多有一个父级,并且可以有多个子级。如果没有做到这一点,或者有多个树没有连接,那么机器人的建模就不正确。按照惯例,有几个重要的轴:

- base_footprint 通常是机器人 TF 树的根节点,对应于机器人底部的中心。将机器人传感器信息转换到这个轴有助于将它们相互关联。
- base_link 通常是 /base_footprint 的子节点,是指机器人的中心,它已经高于地面水平。
- odom 是 /base_footprint 的父坐标系,它们之间的变换表示了机器人自开始运动以来的位移。

图 4.1 展示了模拟 Tiago 机器人的部分 TF 树以及 RViz2 中的 TF 显示。如果你想查看完整的 TF 树,请启动模拟并输入以下命令[1]:

```
$ ros2 run rqt_tf_tree rqt_tf_tree
```

表 4.1 展示了模拟 Tiago 机器人的 TF 树各部分的含义和说明[2]。

表 4.1 模拟 Tiago 机器人的 TF 树各部分的含义和说明

部分	含义	说明
odom	里程计坐标系	指机器人根据自身轮子的转动来估计其在世界坐标系中的位置和方向的坐标系
base_footprint	基础坐标系	指机器人在地面上的接触点或中心点的坐标系
base_link	基础链接坐标系	指机器人的一个主要坐标系,作为其他坐标系的参考
base_cover_link	基础覆盖链接	指机器人某个部分的坐标系
rgbd_laser_link	RGBD 激光器链接	指带有 RGB 颜色和深度信息的激光扫描器的坐标系
base_sonar_03_link	基础声呐 03 链接	指机器人上编号为 03 的声呐传感器的坐标系或数据链接
torso_fixed_column_link	躯干固定柱链接	指机器人躯干上用于固定的柱状结构的坐标系
torso_fixed_link	躯干固定链接	指机器人躯干部分的固定坐标系,用于其他传感器或机械臂的固定参考
torso_lift_link	躯干升降链接	指机器人躯干部分的升降机构的坐标系
head_1_link	头部 1 链接	指机器人头部的第一个传感器或执行器的坐标系
head_2_link	头部 2 链接	指机器人头部的第二个传感器或执行器的坐标系

当一个节点想要使用这个系统时,它不会直接订阅这些主题,而是使用 TF 监听器对象,这些对象会更新一个缓冲区,该缓冲区存储所有最新发布的 TF,并且有一个 API 允许:
- 了解在时间 t 时,从一个坐标系到另一个坐标系是否存在 TF。
- 了解在时间 t 时,从坐标系 A 到坐标系 B 的旋转是什么。

---

[1] 需要安装名为 ros-humble-rqt-tf-tree 的包。

[2] 该对照表为译者注释,原书无此部分。——编辑注

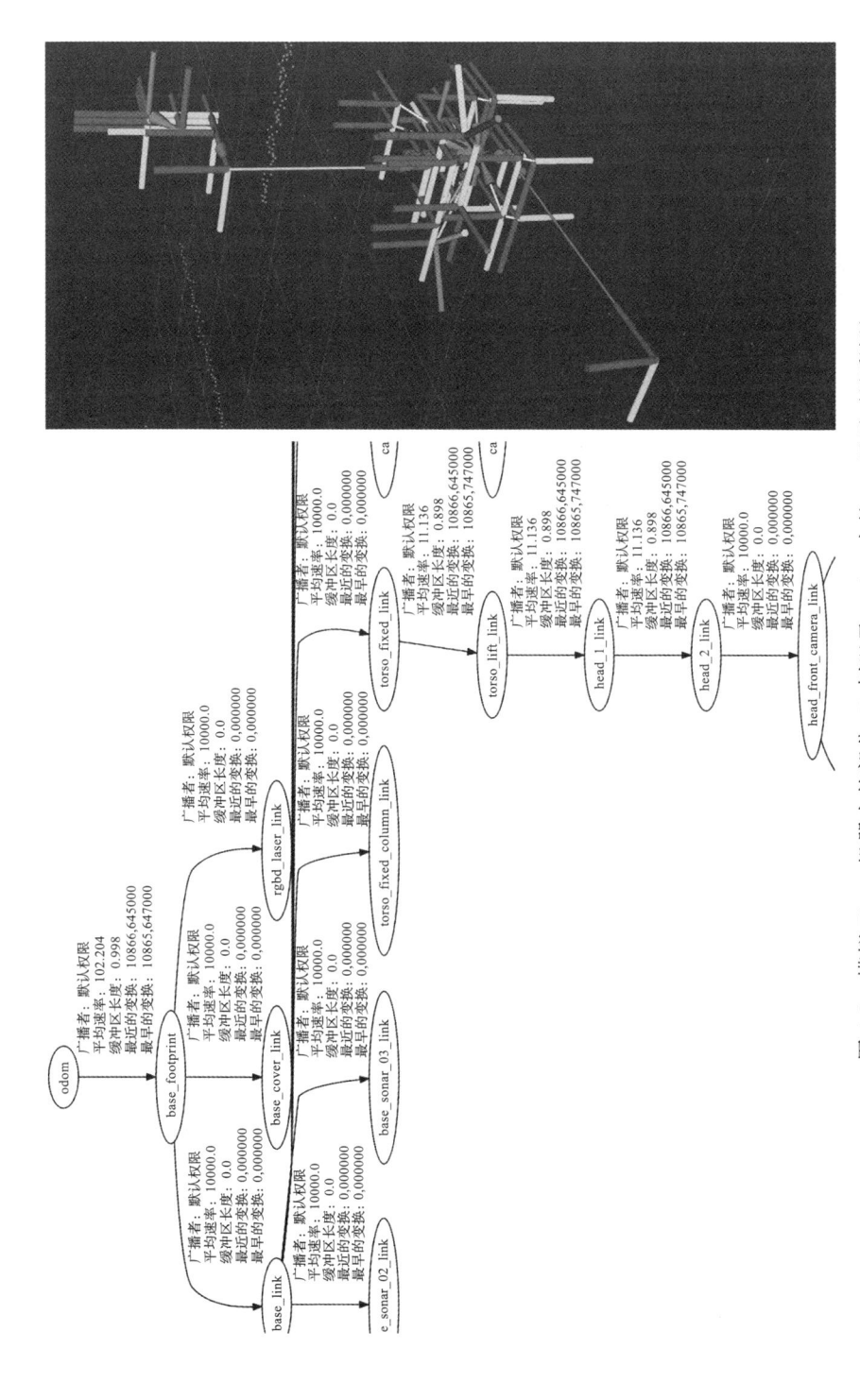

图 4.1 模拟 Tiago 机器人的部分 TF 树以及 RViz2 中的 TF 显示（见彩插）

- 请求在任意时间 t 将坐标系 A 中的坐标转换到坐标系 B。

缓冲区可能不仅包含时间 t 的 TF，如果它有一个更早和一个稍晚的 TF，它还会执行插值。同样，坐标系 A 和 B 可能不是直接连接的，其中间可能还有更多的坐标系，自动执行必要的矩阵运算。

不详细展开细节，目前向 ROS2 节点发布一个变换是非常简单的。只需要有一个变换广播器，并将变换发送到 TF 系统即可：

```
geometry_msgs::msg::TransformStamped detection_tf;

detection_tf.header.frame_id = "base_footprint";
detection_tf.header.stamp = now();
detection_tf.child_frame_id = "detected_obstacle";
detection_tf.transform.translation.x = 1.0;

tf_broadcaster_->sendTransform(detection_tf);
```

获取变换也很简单。通过一个变换监听器更新的 TF 缓冲区，我们可以请求从一个坐标系到另一个坐标系的几何变换，甚至这些坐标系不需要直接连接。任何计算都对开发者透明：

```
tf2_ros::Buffer tfBuffer;
tf2_ros::TransformListener tfListener(tfBuffer);

...

geometry_msgs::msg::TransformStamped odom2obstacle;
odom2obstacle = tfBuffer_.lookupTransform("odom", "detected_obstacle", tf2::TimePointZero);
```

上面的代码自动计算了 odom→base_footprint→detected_obstacle 的变换。lookupTransform 的第三个参数表示我们想要获取变换的时间点。tf2::TimePointZero 表示获取最新的可用变换。例如，如果我们正在转换激光点的坐标，我们就应该使用激光消息头中出现的时间戳，因为如果从那时起机器人或激光移动了，则其他时刻的变换将不会是精确的（在机器人上，几毫秒可能会发生很多变化）。最后，请注意，在使用 now() 请求变换时要小心，因为它在这个时间点还没有信息，也不能被推测到未来，这可能会导致一个异常。

我们可以对变换进行操作，例如对它们进行乘法运算或计算它们的逆变换。接下来，我们将为代码建立一套命名约定。这将有助于我们操作变换 TF：
- 如果一个对象代表从源坐标系到目标坐标系的变换，我们就称之为 origin2target。
- 需要将两个 TF 变换相乘，命名和操作 TF 的助记规则如图 4.2 所示。
  1）只有在运算符 * 附近的坐标系名称相等时，我们才能对其进行操作。在这种情况下，坐标系名称是相同的（robot）。
  2）结果的坐标系 id 必须是运算符的外侧部分（第一个运算符的 odom 和第二个运算符的 object）。
  3）如果我们反转一个 TF（它们是可逆的），我们就在这个名字中反转坐标系 id。

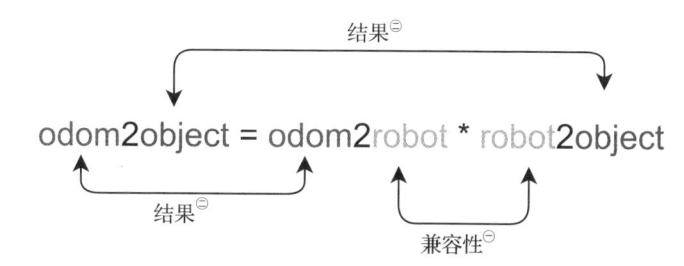

$$\text{odom2object} = \text{odom2robot} * \text{robot2object}$$

（上方箭头：结果①；下方左箭头：结果①；下方右箭头：兼容性①）

图 4.2　命名和操作 TF 的助记规则。根据它们的名称，我们可以知道两个变换 TF 是否可以相乘以及输出结果 TF 的名称

4.1　使用 TF2 的障碍物检测器

本节将分析一个项目，以对前面介绍的 TF 概念进行应用实践。

这个项目使用激光传感器检测机器人正前方的障碍物，如图 4.3 所示。

图 4.3　机器人 Tiago 使用激光传感器检测障碍物。箭头标识中央读数检测到的障碍物（见彩插）

我们将应用 TF 的概念，遵循 ROS2 软件包中的常见实践，将感知结果发布为 TF。这样做的优点是，我们可以轻松地为任何坐标系推理其几何位置，即使它当前没有被感知到。

我们不会引入一个新的感知模型，而是使用与前一章相同的一个模型；我们将

---

① 这里指的是检查或说明 odom2robot 和 robot2object 两个变换是否可以合法组合。——译者注
① odom 到 object 的变换等于 odom 到 robot 的变换 *#robot 到 object 的变换。——译者注

使用激光检测机器人前方的障碍物。使用相同的基于速度的驱动模型，尽管在这种情况下我们会手动遥控机器人。

在这个项目中，除了使用关于 TF 的概念，我们还将展示一个强大的调试工具，即视觉标记○，它允许我们发布可以在 RViz2○中查看的 3D 视觉元素。这种机制允许我们直观地展示机器人的部分内部状态，而不局限于使用 RCLCPP_* 宏生成的调试消息。标记包括箭头、线条、圆柱体、球体、形状、文本等，可以是任意大小或颜色。图 4.4 显示了可用于可视化调试的视觉标记的。

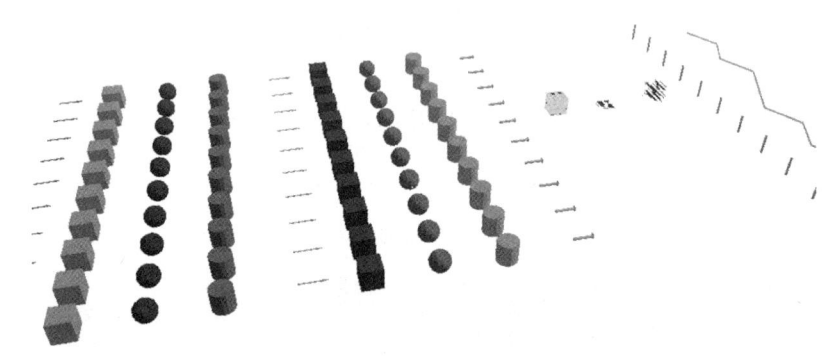

图 4.4 可用于可视化调试的视觉标记（见彩插）

4.2 计算图

我们的应用程序的计算图如图 4.5 所示。

该节点使用模拟机器人的激光传感器，订阅 scan_raw 主题。检测节点订阅激光主题，并在 ROS2 TF 子系统中发布变换。我们的节点订阅 /input_scan，因此我们需要从 /scan_raw 重映射。

我们将创建一个名为 /obstacle_monitor 的节点，该节点读取与检测对应的变换，并在控制台中显示其相对于机器人的 base_footprint 的位置。

/obstacle_monitor 节点还会发布一个视觉标记。在我们的案例中，我们将发布一个红色箭头，它连接机器人的基座与我们所发布的障碍物坐标系的位置。

在本项目中，我们将制作两个版本：一个基本版本和一个改进版本。这样做的原因是要观察使用 TF 的一个小细节，它对最终结果产生了显著的影响，稍后我们将作出解释。

○ http://wiki.ros.org/rviz/DisplayTypes/Marker。
○ 这是一个用于三维可视化的 ROS 工具，用于展示传感器数据、机器人模型等。——译者注

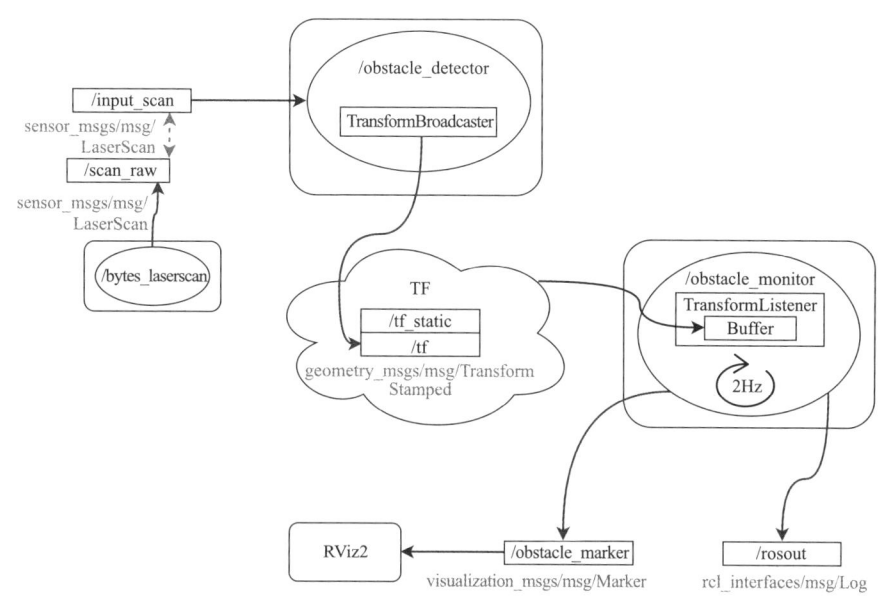

图 4.5　我们的应用程序的计算图。/obstacle_detector 节点与 /obstacle_monitor 节点通过 TF 子系统进行协作

在这个项目中，我们将制作两个版本：一个基本版和一个改进版。这样做的原因是为了观察关于 TF 使用的一个小细节，该细节对最终结果有显著影响，我们将在后面解释。

4.3　基础检测器

我们在这两个版本中使用相同的包。该包的结构可以在下面的方框中看到：

```
Package br2_tf2_detector

br2_tf2_detector
├── CMakeLists.txt
├── include
│   └── br2_tf2_detector
│       ├── ObstacleDetectorImprovedNode.hpp
│       ├── ObstacleDetectorNode.hpp
│       └── ObstacleMonitorNode.hpp
├── launch
│   ├── detector_basic.launch.py
│   └── detector_improved.launch.py
├── package.xml
└── src
    ├── br2_tf2_detector
    │   ├── ObstacleDetectorImprovedNode.cpp
    │   ├── ObstacleDetectorNode.cpp
    │   └── ObstacleMonitorNode.cpp
    ├── detector_improved_main.cpp
    └── detector_main.cpp
```

这里，我们将忽略文件名中包含"Improved"一词的文件。我们将在下一节中看到它们。

读者可以看到包的结构与前一章相似。节点在它们的声明和定义中是分开的，位于与包名称相匹配的目录中。此外，所有内容都将定义在一个与包名称匹配的命名空间中。这个包在这个结构中将迈出一步：现在，我们将节点编译为动态库，它由可执行文件链接。在这个项目中，我们可能不会注意到这个区别，但这样做可以节省空间，也更方便，并且如果需要（在这个项目中并不需要），则可以将其导出到其他包。库的名称将是包的名称（${PROJECT_NAME}），这是在包中创建支持库时的通常做法。让我们看看 CMakeLists.txt 文件的样子：

```
include/br2_tf2_detector/ObstacleDetectorNode.hpp
project(br2_tf2_detector)

find_package(...)
...
)
set(dependencies
...
)

include_directories(include)

add_library(${PROJECT_NAME} SHARED
  src/br2_tf2_detector/ObstacleDetectorNode.cpp
  src/br2_tf2_detector/ObstacleMonitorNode.cpp
  src/br2_tf2_detector/ObstacleDetectorImprovedNode.cpp
)
ament_target_dependencies(${PROJECT_NAME} ${dependencies})

add_executable(detector src/detector_main.cpp)
ament_target_dependencies(detector ${dependencies})
target_link_libraries(detector ${PROJECT_NAME})

add_executable(detector_improved src/detector_improved_main.cpp)
ament_target_dependencies(detector_improved ${dependencies})
target_link_libraries(detector_improved ${PROJECT_NAME})

install(TARGETS
  ${PROJECT_NAME}
  detector
  detector_improved
  ARCHIVE DESTINATION lib
  LIBRARY DESTINATION lib
  RUNTIME DESTINATION lib/${PROJECT_NAME}
)
```

请注意，现在需要添加一个 target_link_libraries 语句，并在与可执行文件相同的位置安装库。在指定每个可执行文件的文件时，不再需要指定超过主 cpp 程序文件之外的文件。

4.3.1 障碍检测节点

分析障碍检测节点。它的执行遵循事件驱动模型，而不是任务迭代模型。节点接收的每个消息都将产生一个输出，因此节点的逻辑位于激光回调中是合理的。

```
include/br2_tf2_detector/ObstacleDetectorNode.hpp
```
```cpp
class ObstacleDetectorNode : public rclcpp::Node
{
public:
  ObstacleDetectorNode();

private:
  void scan_callback(sensor_msgs::msg::LaserScan::UniquePtr msg);

  rclcpp::Subscription<sensor_msgs::msg::LaserScan>::SharedPtr scan_sub_;
  std::shared_ptr<tf2_ros::StaticTransformBroadcaster> tf_broadcaster_;
};
```

由于节点必须向 TF 子系统发布变换，因此我们声明了一个 StaticTransformBroadcaster，它发布在 /tf_static 消息主题。我们也可以声明一个 TransformBroadcaster，它发布在 /tf 消息主题。除了持久性 QoS 之外，两者的区别在于我们希望变换持续超过默认的 10s，而不仅仅是非静态变换。

我们使用一个 shared_ptr 共享指针来管理 tf_broadcaster_，因为它的构造函数需要一个 rclcpp::Node* 类型的参数。而我们在构造函数内部才可以获取到它[⊖]：

```
src/br2_tf2_detector/ObstacleDetectorNode.hpp
```
```cpp
ObstacleDetectorNode::ObstacleDetectorNode()
: Node("obstacle_detector")
{
  scan_sub_ = create_subscription<sensor_msgs::msg::LaserScan>(
    "input_scan", rclcpp::SensorDataQoS(),
    std::bind(&ObstacleDetectorNode::scan_callback, this, _1));

  tf_broadcaster_ = std::make_shared<tf2_ros::TransformBroadcaster>(*this);
}
```

tf_broadcaster_object 负责发布静态 TF。TF 的消息类型是 geometry_msgs/msg/TransformStamped。让我们看看它的使用方法：

```
src/br2_tf2_detector/ObstacleDetectorNode.hpp
```
```cpp
void
ObstacleDetectorNode::scan_callback(sensor_msgs::msg::LaserScan::UniquePtr msg)
{
  double dist = msg->ranges[msg->ranges.size() / 2];

  if (!std::isinf(dist)) {
    geometry_msgs::msg::TransformStamped detection_tf;

    detection_tf.header = msg->header;
    detection_tf.child_frame_id = "detected_obstacle";
    detection_tf.transform.translation.x = msg->ranges[msg->ranges.size() / 2];

    tf_broadcaster_->sendTransform(detection_tf);
  }
}
```

⊖ 实际上，一些 C++ 开发者建议避免在构造函数中使用 this，因为直到构造函数完成，对象才被完全初始化。

- 输出消息的消息头将是输入激光消息的消息头。我们这样做是因为时间戳必须是感测读数取得的时间。如果我们使用 now()，那么根据消息的延迟和计算机的负载，变换可能不会精确，也可能会发生同步错误。

  frame_id 是变换的源坐标系（或父坐标系），已经在这个头部中。在这种情况下，它是传感器坐标系，因为感知的对象坐标就在这个坐标系中。

- child_frame_id 字段是我们将要创建的新坐标系的 id，它代表感知到的障碍物。
- transfrom 字段包含了从父坐标系到我们要创建的子坐标系的平移和旋转。由于激光坐标系的 x 轴与我们要测量的激光束对齐，因此 x 轴的平移是读取的距离。旋转指的是在平移应用后坐标系的旋转。由于这个值在这里不相关（这里检测的是一个点），因此我们使用消息构造函数设置的默认四元数值（0，0，0，1）。
- 最后，使用 tf_broadcaster_ 的 sendTransform() 方法将变换发送到 TF 子系统。

4.3.2　障碍物监控节点

/obstacle_monitor 节点从 TF 系统中提取检测到的对象的变换，并以两种方式向用户显示：

- 控制台的标准输出始终指示障碍物相对于机器人的位置，即使它不再被检测到。
- 使用一个视觉标记（具体是一个箭头），从机器人指向检测到的障碍物。

分析头部以查看该节点包含哪些元素：

```
include/br2_tf2_detector/ObstacleMonitorNode.hpp
class ObstacleMonitorNode : public rclcpp::Node
{
public:
  ObstacleMonitorNode();

private:
  void control_cycle();
  rclcpp::TimerBase::SharedPtr timer_;

  tf2::BufferCore tf_buffer_;
  tf2_ros::TransformListener tf_listener_;

  rclcpp::Publisher<visualization_msgs::msg::Marker>::SharedPtr marker_pub_;
};
```

- 该节点的执行模型是迭代式的，所以我们声明了一个（定时器）及其回调 control_cycle。
- 要访问 TF 系统，我们使用一个 tf2_ros::TransformListener，它会更新 tf_buffer_ 缓存区，我们可以从中进行所需的查询。
- 对于视觉标记，我们只需要一个发布者。

在类的定义中，我们暂时忽略与视觉标记相关的部分，现在只展示与 TF 相关的部分。

```
src/br2 tf2 detector/ObstacleMonitorNode.cpp
1   ObstacleMonitorNode::ObstacleMonitorNode()
2   : Node("obstacle_monitor"),
3     tf_buffer_(),
4     tf_listener_(tf_buffer_)
5   {
6     marker_pub_ = create_publisher<visualization_msgs::msg::Marker>(
7       "obstacle_marker", 1);
8
9     timer_ = create_wall_timer(
10      500ms, std::bind(&ObstacleMonitorNode::control_cycle, this));
11  }
12
13  void
14  ObstacleMonitorNode::control_cycle()
15  {
16    geometry_msgs::msg::TransformStamped robot2obstacle;
17
18    try {
19      robot2obstacle = tf_buffer_.lookupTransform(
20        "base_footprint", "detected_obstacle", tf2::TimePointZero);
21    } catch (tf2::TransformException & ex) {
22      RCLCPP_WARN(get_logger(), "Obstacle transform not found: %s", ex.what());
23      return;
24    }
25
26    double x = robot2obstacle.transform.translation.x;
27    double y = robot2obstacle.transform.translation.y;
28    double z = robot2obstacle.transform.translation.z;
29    double theta = atan2(y, x);
30
31    RCLCPP_INFO(get_logger(), "Obstacle detected at (%lf m, %lf m, , %lf m) = %lf
32      rads", x, y, z, theta);
33  }
```

- 注意 tf_listener_ 是如何通过简单指定要更新的缓冲区来初始化的。稍后，查询将直接对缓冲区进行。
- 我们观察到控制循环以 2Hz 的频率运行，并通过 RCLCPP_INFO（输出到 /ros_out 和 stdout）显示信息。
- 最相关的函数是 lookupTransform，它计算从一个坐标系到另一个坐标系的几何变换，即使它们之间没有直接的关联。我们可以指定一个特定的时间戳，或者，相反，如果我们想要最新的可用变换，则可以通过指定 tf2::TimePointZero 来实现。如果它不存在，这个调用可能会抛出一个异常，或者我们需要一个在未来时间戳的变换，所以应该使用 try/catch 来处理可能的错误。
- 请注意，我们在 ObstacleDetectorNode 中发布的 TF 是 base_laser_link→detected_obstacle 转换，而现在我们需要的 TF 是 base_footprint→detected_obstacle 转换。由于机器人被很好地建模，base_laser_link 和 base_footprint 之间的几何关系可以被计算，因此 lookupTransform 将能够返回正确的信息。

让我们看看与生成视觉标记相关的部分。目标是将障碍物的坐标显示在机器人的屏幕上，并在 RViz2 中显示一个几何形状，以便我们能够直观地调试应用程序。

在这种情况下，它将是一个从机器人到障碍物的红色箭头。为此，我们需要创建一个 visualization_msgs/msg/Marker 消息，并填写其字段以获得这个箭头：

```
src/br2_tf2_detector/ObstacleMonitorNode.cpp
visualization_msgs::msg::Marker obstacle_arrow;
obstacle_arrow.header.frame_id = "base_footprint";
obstacle_arrow.header.stamp = now();
obstacle_arrow.type = visualization_msgs::msg::Marker::ARROW;
obstacle_arrow.action = visualization_msgs::msg::Marker::ADD;
obstacle_arrow.lifetime = rclcpp::Duration(1s);

geometry_msgs::msg::Point start;
start.x = 0.0;
start.y = 0.0;
start.z = 0.0;
geometry_msgs::msg::Point end;
end.x = x;
end.y = y;
end.z = z;
obstacle_arrow.points = {start, end};

obstacle_arrow.color.r = 1.0;
obstacle_arrow.color.g = 0.0;
obstacle_arrow.color.b = 0.0;
obstacle_arrow.color.a = 1.0;

obstacle_arrow.scale.x = 0.02;
obstacle_arrow.scale.y = 0.1;
obstacle_arrow.scale.z = 0.1;
```

在视觉标记的参考文档中，记录了每种类型的标记的每个字段的含义。对于箭头，我们需要填充 points 字段，起点是 (0，0，0)，终点是检测到的位置，它们都在 base_footprint 坐标系中。不要忘记分配颜色，尤其是 alpha 值，因为如果将其设置为 0（默认值），我们将看不到任何东西。

4.3.3 运行基本检测器

我们在同一进程中实例化两个节点来测试我们的节点，并且使用 Single-ThreadedExecutor 单线程执行器。这已足够让这两个节点同时运作：

```
src/br2_tf2_detector/ObstacleMonitorNode.cpp
int main(int argc, char * argv[]) {
  rclcpp::init(argc, argv);

  auto obstacle_detector = std::make_shared<br2_tf2_detector::ObstacleDetectorNode>();
  auto obstacle_monitor = std::make_shared<br2_tf2_detector::ObstacleMonitorNode>();

  rclcpp::executors::SingleThreadedExecutor executor;
  executor.add_node(obstacle_detector->get_node_base_interface());
  executor.add_node(obstacle_monitor->get_node_base_interface());

  executor.spin();

  rclcpp::shutdown();
  return 0;
}
```

请按照以下步骤测试我们的节点：

```
# Terminal 1: The Tiago simulation
$ ros2 launch br2_tiago sim.launch.py world:=empty

# Terminal 2: Launch our nodes
$ ros2 launch br2_tf2_detector detector_basic.launch.py

# Terminal 3: Keyboard teleoperation
$ ros2 run teleop_twist_keyboard teleop_twist_keyboard --ros-args -r
cmd_vel:=/key_vel

# Terminal 4: RViz2
$ ros2 run rviz2 rviz2 --ros-args -p use_sim_time:=true
```

在 Gazebo 中，在机器人前方添加一个障碍物。在终端中查看检测到的信息。在 RViz2 中，将固定坐标系更改为 odom。在 RViz2 中添加一个标记显示，指定为我们创建的用于发布视觉标记的主题。如果尚未添加 TF 显示，则需添加它。图 4.6 展示了 RViz2 中检测到的 TF 的视觉表示，以及红色箭头标记。

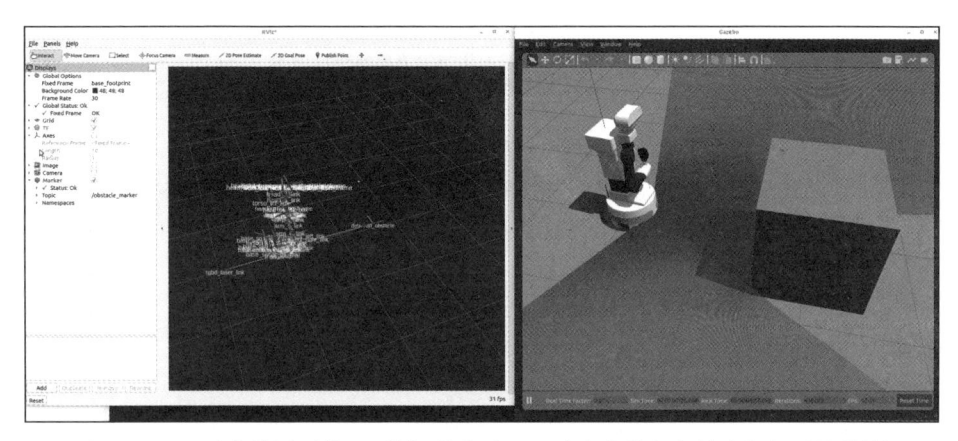

图 4.6　RViz2 中检测到的 TF 的视觉表示，以及发布的红色箭头标记（见彩插）

进行一个快速的练习：使用遥控器，让机器人向前和向侧移动，使其不再感知到障碍物。继续移动机器人，注意到 lookupTransform 返回的信息不再正确。它继续指示障碍物在前面，尽管这已经不再正确。这是发生了什么呢？我们可能希望箭头指向障碍物的位置，但现在箭头固定在机器人上。

图 4.7 展示了在本地坐标系中发布 TF 时的问题。只要机器人感知到障碍物，请求的变换（粉红色箭头）就是正确的。这是一个从机器人的激光到障碍物的变换。当我们因障碍物已经消失而停止更新变换（粗蓝色箭头）时，该变换仍然存在。如

果我们移动机器人，lookupTransform 就继续返回最后一个有效的变换：在机器人前面。这使得视觉标记也出现了错误。接下来的部分将介绍一种策略来解决这个问题。

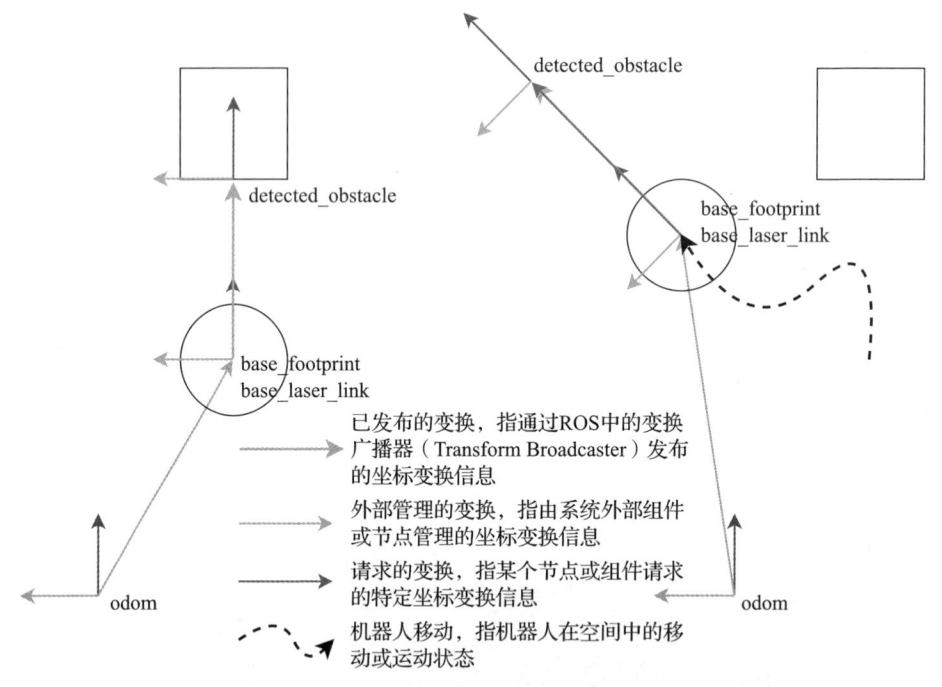

图 4.7 展示在本地坐标系中发布 TF 时的问题。当机器人移动时，TF 不再代表正确的障碍物位置（见彩插）

4.4 改进的检测器

解决方案是在一个固定的坐标系中发布检测 TF，这个坐标系不会受到机器人的移动的影响，例如，odom（如果你的机器人正在导航，则可以用 map）。如果我们这样做，当我们需要变换 base_footprint→detected_obstacle（粉色箭头）时，这个变换将考虑到机器人的移动，这在变换 odom→base_footprint 中收集。图 4.8 展示了如何通过在固定帧中发布 TF 来正确维护障碍物位置。

ObstacleDetectorImprovedNode 是对 ObstacleDetectorNode 的修改，以实现这个改进。这个新节点操作 TF，在某些时候，它会查询一个现有 TF 的值。因此，除了拥有一个 StaticTransformPublisher 外，它还实例化了一个带有相关 Buffer 的 TransformListener。

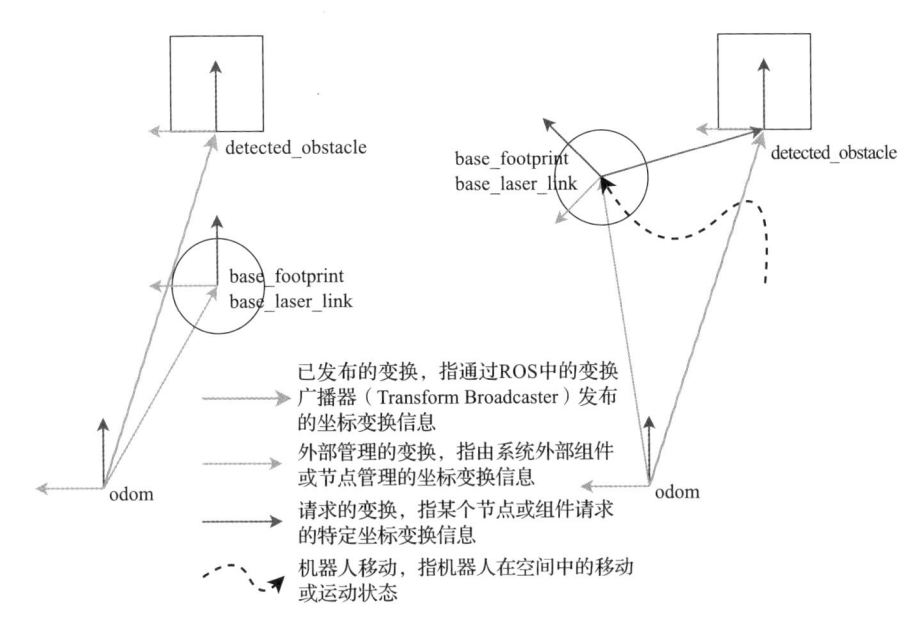

图 4.8 展示了如何通过在固定坐标系中发布 TF 来正确维护障碍物位置。
计算的 TF（粗蓝箭头）考虑了机器人的位移（见彩插）

```
include/br2_tf2_detector/ObstacleMonitorNode.hpp
class ObstacleDetectorImprovedNode : public rclcpp::Node
{
...
private:
  ...
  tf2::BufferCore tf_buffer_;
  tf2_ros::TransformListener tf_listener_;
};
```

在实现这个节点时，检查两个相关但不同的数据结构：

- geometry_msgs::msg::TransformStamped 是一种消息类型，用于发布 TF，并且是 lookupTransform 的返回结果。
- tf2::Transform 是 TF2 库的一种数据类型，允许进行操作。
- tf2::Stamped<tf2::Transform> 与前者类似，但带有一个指示时间戳的头部。它将需要符合变换函数中的类型。
- tf2::fromMsg/tf2::toMsg 是变换函数，允许从消息类型变换到 TF2 类型，反之亦然。

作为一个普遍的建议，不要在节点内部使用消息类型来操作它们。将这个建议应用于 TF、图像、点云⊖等数据类型。消息是节点间通信的好工具，但功能非常有限。如果

⊖ 点云（Point Cloud）是一个由大量点组成的数据集，这些点在三维空间中代表物体的外部表面或某些测量结果。——译者注

有一个库提供了本地类型，就使用它，因为它将更有用。通常，有函数可以从消息类型转换到本地类型。在这种情况下，我们使用 geometry_msgs::msg::TransformStamped 来发送和接收 TF，但使用 TF2 库来操作它们。

考虑到之前建立的约定，让我们看看如何进行改进。如前所述，我们的目标是创建 TF 的 odom2object（object 是检测到的障碍物）。观察值表示为变换 laser2object，所以我们必须在下面的等式中找到 X：

$$\text{odom2object} = X * \text{laser2object}$$

通过从我们之前陈述的规则推断，X 必须为 odom2laser，这是一个可以从 lookupTransform 请求的 TF。

```cpp
include/br2_tf2_detector/ObstacleMonitorNode.hpp

double dist = msg->ranges[msg->ranges.size() / 2];
if (!std::isinf(dist)) {
  tf2::Transform laser2object;
  laser2object.setOrigin(tf2::Vector3(dist, 0.0, 0.0));
  laser2object.setRotation(tf2::Quaternion(0.0, 0.0, 0.0, 1.0));

  geometry_msgs::msg::TransformStamped odom2laser_msg;
  tf2::Stamped<tf2::Transform> odom2laser;
  try {
    odom2laser_msg = tf_buffer_.lookupTransform(
      "odom", "base_laser_link", msg->header.stamp, rclcpp::Duration(200ms));
    tf2::fromMsg(odom2laser_msg, odom2laser);
  } catch (tf2::TransformException & ex) {
    RCLCPP_WARN(get_logger(), "Obstacle transform not found: %s", ex.what());
    return;
  }

  tf2::Transform odom2object = odom2laser * laser2object;

  geometry_msgs::msg::TransformStamped odom2object_msg;
  odom2object_msg.transform = tf2::toMsg(odom2object);

  odom2object_msg.header.stamp = msg->header.stamp;
  odom2object_msg.header.frame_id = "odom";
  odom2object_msg.child_frame_id = "detected_obstacle";

  tf_broadcaster_->sendTransform(odom2object_msg);
}
```

- laser2object 存储对检测到的对象的感知。它只是在 x 轴上的平移，对应于到障碍物的距离。
- 要获取 odom2laser，我们需要用 lookupTransform 查询 TF 子系统，将结果变换消息转换为需要的类型以操作变换。
- 在这一点上，我们有了计算 odom2object 的所有东西，也就是，障碍物相对于固定坐标系 odom 的位置。
- 最后，我们组合输出消息并发布到 TF 子系统。

不需要对 ObstacleMonitorNode 进行任何更改，因为 lookupTransform 将计算 TF base_footprint→obstacle，TF 系统知道 TF odom→base_footprint 和 odom→obstacle。

执行新的、改进的节点的过程与基本情况类似，唯一的区别是我们在主程序和启动器中指定了改进节点。由于这是一个简单的更改，因此我们将在这里跳过显示。让我们按照类似的命令来执行它：

```
# Terminal 1: The Tiago simulation
$ ros2 launch br2_tiago sim.launch.py world:=empty
```

```
# Terminal 2: Launch our nodes
$ ros2 launch br2_tf2_detector detector_improved.launch.py
```

```
# Terminal 3: Keyboard teleoperation
$ ros2 run teleop_twist_keyboard teleop_twist_keyboard --ros-args -r
cmd_vel:=/key_vel
```

```
# Terminal 4: RViz2
$ ros2 run rviz2 rviz2 --ros-args -p use_sim_time:=true
```

在 Gazebo 中添加障碍物，以便机器人可以检测到它。观察控制台输出和 RViz2 中的视觉标记。移动机器人使障碍物不被检测到，并看到标记和输出现在是正确的。位移（编码为变换 odom→base_footprint）用于正确更新信息。

建议的练习

1）制作一个节点来显示每秒机器人移动了多少次。你可以通过保存 (odom→base_footprint)$_t$，并从 (odom→base_footprint)$_{t+1}$ 中减去它来做到这一点。

2）在 ObstacleDetectorNode 中，根据障碍物的距离改变箭头的颜色：绿色代表远，红色代表近。

3）在 ObstacleDetectorNode 中，在终端中显示障碍物在 odom、base_footprint 和 head_2_link 坐标系中的位置。

CHAPTER 5

第 5 章

反应式行为

反应式行为（reactive behavior）紧密地将感知与行动联系在一起，而不需要使用抽象表示。正如 Brooks 在他的 Subsumption 架构[1]中展示的那样，可以通过激活或抑制更高层次的简单反应行为来创建相对复杂的行为。

本章将不讨论 Subsumption 架构。读者可以参考 Cascade Lifecycle⊖包和 rqt_cascade_hfsm⊜来了解构建 Subsumption 架构的构建块。本章的目的是展示使用不同资源推进 ROS2 知识的几种反应式行为。

本章首先将介绍一个名为虚拟力场（Virtual Force Field，VFF）的简单本地导航算法，它利用激光来避免障碍物。这个例子将建立关于视觉标记的知识，并介绍测试驱动开发方法。其次，本章将介绍基于相机信息的反应式跟踪行为，以及如何处理图像和如何控制机器人的关节。最后，本章还将介绍一种有利的节点类型，称为生命周期节点（Lifecycle Node）。

5.1 使用 VFF 避免障碍物

本节将展示如何实现一个简单的反应式行为，使 Tiago 机器人可通过使用简单的 VFF 算法向前移动并避开障碍物。这个简单的算法使用以下三个二维向量来计算期望速度（或指令速度），从而实现对机器人运动的控制。

⊖ https://github.com/fmrico/cascade_lifecycle。
⊜ https://github.com/fmrico/rqt_cascade_hfsm。

- **吸引向量**：该向量总是指向目标方向（通常是前方），因为机器人能够在无障碍物的情况下直线移动。
- **排斥向量**：该向量是基于激光传感器的读数计算得出的。在基本版本中，距离机器人最近的障碍物会产生一个与距离成反比的排斥向量。
- **结果向量**：该向量是前两个向量的和，它用于计算机器人的期望速度。线速度取决于产生的向量的模量，而转弯的角度取决于产生的向量角度。

图 5.1 展示了由相同障碍物产生的 VFF 向量的示例。

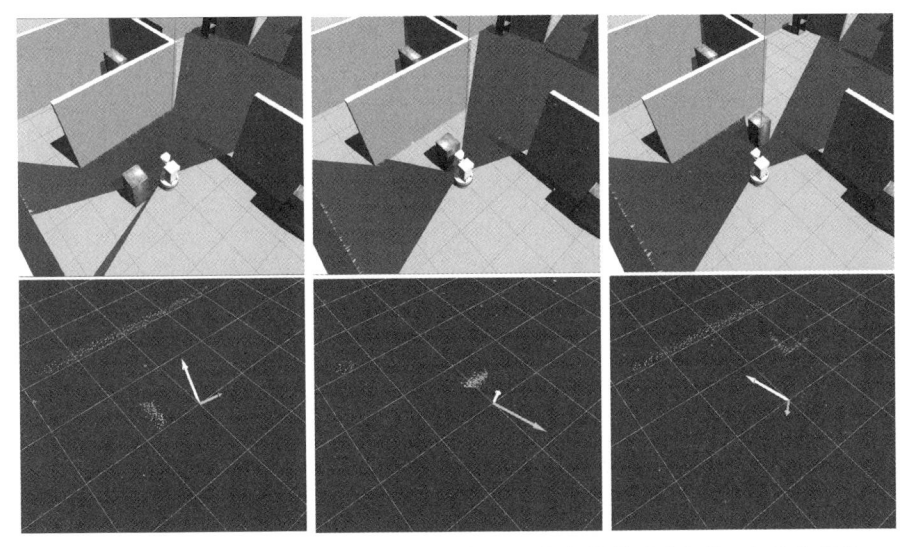

图 5.1　由相同障碍物产生的 VFF 向量示例。蓝色矢量是吸引向量，红色向量是排斥向量，绿色向量是结果矢量（见彩插）

5.1.1　计算图

首先，看一下机器人避障行为的控制算法的计算图是什么样子的。如图 5.2 所示，我们在一个进程中有一个单节点，它具有以下元素和特点：

- 该节点订阅包含感知信息的消息主题，并将其发布到速度消息主题。这些将是主要的输入和输出主题。我们将使用通用名称来指代这些主题，它们在部署时将被重映射。
- 有足够多的信息来确定机器人为什么会以某种方式行动是至关重要的。ROS2 提供了许多调试工具。使用 /rosout 是一个很好的选择。使用机器人配备的 LED 也很方便。有了一个可以改变颜色的 LED，就可以用颜色编码机器人的状态或感知。只要看一眼，我们就能得到很多关于机器人为什么做出决定的信息。在这种情况下,除了上面的输入和输出主题之外,我们还添加了 /vff_debug 调试主题,它发布可视化标记以可视化 VFF 的不同向量。图 5.1 中的颜色向量是节点发布的可视化标记,在 RViz2 中被可视化。
- 在这种情况下,我们将选择由节点使用定时器进行内部控制的迭代执行,并控制频率为 20Hz。

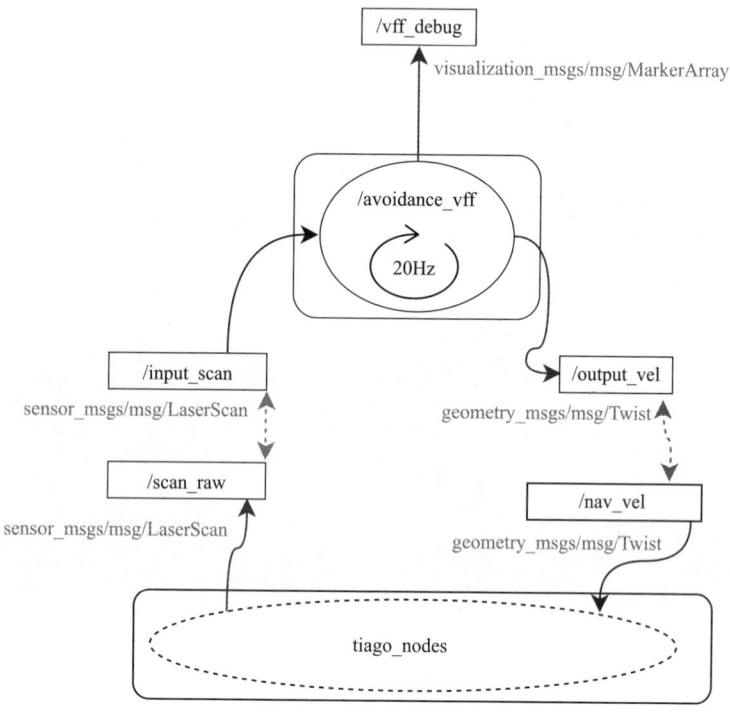

图 5.2　机器人避障行为的控制算法的计算图

5.1.2　软件包结构

下面方框中的软件包结构已成为软件包的建议标准:每个节点都有其对应的声明和定义,分别位于各自的 .hpp 文件和 .cpp 文件中,还有一个将实例化它的主程序。我们有一个启动目录,其中包含启动器,以便轻松运行我们的项目。请注意,我们现在已经添加了一个 tests 目录,其中将包含我们的测试文件,这将在后面解释。

```
Package br2_vff_avoidance

br2_vff_avoidance
├── CMakeLists.txt
├── include
```

```
Package br2_vff_avoidance
    ├── br2_vff_avoidance
    │   └── AvoidanceNode.hpp
    ├── launch
    │   └── avoidance_vff.launch.py
    ├── package.xml
    ├── src
    │   ├── avoidance_vff_main.cpp
    │   └── br2_vff_avoidance
    │       └── AvoidanceNode.cpp
    └── tests
        ├── CMakeLists.txt
        └── vff_test.cpp
```

5.1.3 控制逻辑

避障节点（AvoidanceNode）实现了 VFF 算法，基于激光器读数生成控制命令，它的主要元素与先前的示例类似：

- 一个激光器读数的订阅者，其功能将是在 last_scan_ 中更新最后的读数。
- 一个用于速度的发布者。
- 一个 get_vff 函数，该函数用于根据激光器读数计算 VFF 算法基于的三个向量。我们声明了一个新的类型（VFFVectors）来打包它们。
- 由于这个节点是迭代执行的，因此我们使用一个定时器并使用 control_cycle 方法作为回调函数。

```cpp
// include/br2_vff_avoidance/AvoidanceNode.hpp
struct VFFVectors
{
  std::vector<float> attractive;
  std::vector<float> repulsive;
  std::vector<float> result;
};

class AvoidanceNode : public rclcpp::Node
{
public:
  AvoidanceNode();

  void scan_callback(sensor_msgs::msg::LaserScan::UniquePtr msg);
  void control_cycle();

protected:
  VFFVectors get_vff(const sensor_msgs::msg::LaserScan & scan);

private:
  rclcpp::Publisher<geometry_msgs::msg::Twist>::SharedPtr vel_pub_;
  rclcpp::Subscription<sensor_msgs::msg::LaserScan>::SharedPtr scan_sub_;
  rclcpp::TimerBase::SharedPtr timer_;

  sensor_msgs::msg::LaserScan::UniquePtr last_scan_;
};
```

在控制周期中，首先检查激光器是否有新数据。如果没有，或者这些数据是旧的（如果在过去一秒钟内没有从激光器接收到信息），则不生成控制命令。如果机器人驱动程序正确实现，并且在停止接收命令时不移动，则机器人应该停止移动。在相反的情况下（不适用于我们的案例），你应该发送所有字段为 0 的速度来停止机器人的移动。

计算出结果向量后，通过计算模和角度可以直接将其转换为速度。使用 std::clamp 来控制速度范围处于安全范围内是方便的，如下面的代码所示：

```cpp
// src/br2_vff_avoidance/AvoidanceNode.cpp
void
AvoidanceNode::scan_callback(sensor_msgs::msg::LaserScan::UniquePtr msg)
{
  last_scan_ = std::move(msg);
}

void
AvoidanceNode::control_cycle()
{
  // Skip cycle if no valid recent scan available
  if (last_scan_ == nullptr || (now() - last_scan_->header.stamp) > 1s) {
    return;
  }

  // Get VFF vectors
  const VFFVectors & vff = get_vff(*last_scan_);

  // Use result vector to calculate output speed
  const auto & v = vff.result;
  double angle = atan2(v[1], v[0]);
  double module = sqrt(v[0] * v[0] + v[1] * v[1]);

  // Create ouput message, controlling speed limits
  geometry_msgs::msg::Twist vel;
  vel.linear.x = std::clamp(module, 0.0, 0.3);    // linear vel to [0.0, 0.3] m/s
  vel.angular.z = std::clamp(angle, -0.5, 0.5);   // rotation vel to [-0.5, 0.5] rad/s

  vel_pub_->publish(vel);
}
```

5.1.4 VFF 向量的计算

get_vff 函数的目标是获取三个向量（吸引向量、排斥向量和结果向量）：

```cpp
// src/br2_vff_avoidance/AvoidanceNode.cpp
VFFVectors
AvoidanceNode::get_vff(const sensor_msgs::msg::LaserScan & scan)
{
  // This is the obstacle radius in which an obstacle affects the robot
  const float OBSTACLE_DISTANCE = 1.0;

  // Init vectors
  VFFVectors vff_vector;
  vff_vector.attractive = {OBSTACLE_DISTANCE, 0.0};  // Robot wants to go forward
  vff_vector.repulsive = {0.0, 0.0};
  vff_vector.result = {1.0, 0.0};

  // Get the index of nearest obstacle
  int min_idx = std::min_element(scan.ranges.begin(), scan.ranges.end())
    - scan.ranges.begin();
```

```cpp
src/br2_vff_avoidance/AvoidanceNode.cpp
// Get the distance to nearest obstacle
float distance_min = scan.ranges[min_idx];

// If the obstacle is in the area that affects the robot, calculate repulsive vector
if (distance_min < OBSTACLE_DISTANCE) {
  float angle = scan.angle_min + scan.angle_increment * min_idx;

  float oposite_angle = angle + M_PI;
  // The module of the vector is inverse to the distance to the obstacle
  float complementary_dist = OBSTACLE_DISTANCE - distance_min;

  // Get cartesian (x, y) components from polar (angle, distance)
  vff_vector.repulsive[0] = cos(oposite_angle) * complementary_dist;
  vff_vector.repulsive[1] = sin(oposite_angle) * complementary_dist;
}

// Calculate resulting vector adding attractive and repulsive vectors
vff_vector.result[0] = (vff_vector.repulsive[0] + vff_vector.attractive[0]);
vff_vector.result[1] = (vff_vector.repulsive[1] + vff_vector.attractive[1]);

return vff_vector;
}
```

- 吸引向量始终为（1，0），因为机器人始终尝试向前移动。在假定附近没有障碍物的情况下初始化其余向量。
- 排斥向量是根据最低的激光器读数计算的。通过计算最小值索引（min_idx）为具有较小值的向量的索引，我们能够得到距离（在范围向量中的值）和角度（从 angle_min、angle_increment 和 min_idx 三个变量计算而来）。
- 排斥向量的幅度必须与障碍物距离成反比。更近的障碍物必须比更远的障碍物产生更多的排斥力。
- 排斥向量的角度必须与检测到的障碍物角度相反，因此排斥向量的角度要加上 π。
- 在计算排斥向量的笛卡儿坐标后，我们将其与吸引向量相加以获得其结果向量。

5.1.5 使用视觉标记进行调试

在前一章中，我们使用视觉标记来可视化地调试机器人的行为。图 5.1 中的箭头是由 AvoidanceNode 生成的用于调试的可视化标记。区别在于使用的是 visualization_msgs::msg::MarkerArray 而不是 visualization_msgs::msg::Marker。基本上，一个 visualization_msgs::msg::MarkerArray 在其标记字段中包含了一个 visualization_msgs::msg::Marker 的 std::vector。让我们看看将作为调试信息发布的消息是如何组成的。有关这些消息的详细信息，请查看消息定义和参考页面：

```
$ ros2 interface show visualization_msgs/msg/MarkerArray
Marker[] markers

$ ros2 interface show visualization_msgs/msg/Marker
```

AvoidanceNode 头文件中包含了需要组成和发布可视化标记的内容。我们定义了一个消息的发布者 visualization_msgs::msg::MarkerArray 变量和两个函数，这可以帮助我们构造向量。get_debug_vff 返回由表示三个向量的三个箭头组成的完整消息。

为避免在此函数中重复使用代码，make_marker 创建具有指定颜色的标记作为输入参数。

```
include/br2_vff_avoidance/AvoidanceNode.hpp
typedef enum {RED, GREEN, BLUE, NUM_COLORS} VFFColor;

class AvoidanceNode : public rclcpp::Node
{
public:
  AvoidanceNode();

protected:
  visualization_msgs::msg::MarkerArray get_debug_vff(const VFFVectors & vff_vectors);
  visualization_msgs::msg::Marker make_marker(
    const std::vector<float> & vector, VFFColor vff_color);

private:
  rclcpp::Publisher<visualization_msgs::msg::MarkerArray>::SharedPtr vff_debug_pub_;
};
```

只要有对该信息感兴趣的订阅者（在本例中是 RViz2），标记就被发布到 control_cycle。

```
void
AvoidanceNode::control_cycle()
{
  // Get VFF vectors
  const VFFVectors & vff = get_vff(*last_scan_);

  // Produce debug information, if any interested
  if (vff_debug_pub_->get_subscription_count() > 0) {
    vff_debug_pub_->publish(get_debug_vff(vff));
  }
}
```

对于每个向量，使用不同的颜色创建一个 visualization_msgs::msg::MarkerArray。base_footprint 是在地面上的坐标系，在机器人的中心，面向目标方（通常是前方）。在这个坐标系中，箭头的原点是（0，0），箭头的端点是每个向量所表示的。每个向量必须有一个不同的 id，因为在 RViz2 中，一个标记会用相同的 id 替换另一个标记。

```
visualization_msgs::msg::MarkerArray
AvoidanceNode::get_debug_vff(const VFFVectors & vff_vectors)
{
  visualization_msgs::msg::MarkerArray marker_array;

  marker_array.markers.push_back(make_marker(vff_vectors.attractive, BLUE));
  marker_array.markers.push_back(make_marker(vff_vectors.repulsive, RED));
  marker_array.markers.push_back(make_marker(vff_vectors.result, GREEN));

  return marker_array;
}
```

```cpp
visualization_msgs::msg::Marker
AvoidanceNode::make_marker(const std::vector<float> & vector, VFFColor vff_color)
{
  visualization_msgs::msg::Marker marker;

  marker.header.frame_id = "base_footprint";
  marker.header.stamp = now();
  marker.type = visualization_msgs::msg::Marker::ARROW;
  marker.id = visualization_msgs::msg::Marker::ADD;

  geometry_msgs::msg::Point start;
  start.x = 0.0;
  start.y = 0.0;
  geometry_msgs::msg::Point end;
  start.x = vector[0];
  start.y = vector[1];
  marker.points = {end, start};

  marker.scale.x = 0.05;
  marker.scale.y = 0.1;

  switch (vff_color) {
    case RED:
      marker.id = 0;
      marker.color.r = 1.0;
      break;
    case GREEN:
      marker.id = 1;
      marker.color.g = 1.0;
      break;
    case BLUE:
      marker.id = 2;
      marker.color.b = 1.0;
      break;
  }
  marker.color.a = 1.0;

  return marker;
}
```

5.1.6 运行 AvoidanceNode

运行这个节点的主程序现在对读者来说应该很简单了，只需要实例化节点并调用它来旋转（spin）：

```cpp
// src/avoidance_vff_main.cpp

int main(int argc, char * argv[])
{
  rclcpp::init(argc, argv);

  auto avoidance_node = std::make_shared<br2_reactive_behaviors::AvoidanceNode>();
  rclcpp::spin(avoidance_node);
  rclcpp::shutdown();

  return 0;
}
```

要运行此节点，我们必须首先运行模拟器：

```
$ ros2 launch mr2_tiago sim.launch.py
```

接下来，执行节点设置重映射和参数：

```
$ ros2 run br2_vff_avoidance avoidance_vff --ros-args -r input_scan:=/scan_raw -r
output_vel:=/key_vel -p use_sim_time:=true
```

或使用启动器：

```
$ ros2 launch br2_vff_avoidance avoidance_vff.launch.py
```

如果一切顺利，机器人就开始向前移动。使用模拟器中的按钮移动物体，为机器人设置障碍物。打开 RViz2，并添加类型为 visualization_msgs::msg::MarkerArray 的 /vff_debug 主题的可视化，避障行为的执行如图 5.3 所示。看看节点标记的视觉信息如何帮助我们更好地理解机器人正在做什么。

图 5.3 避障行为的执行（见彩插）

5.1.7 开发过程中的测试

在前面的部分中展示的代码可能包含一些计算错误，这些错误可以在将其在真正的机器人甚至模拟器上运行之前被检测到。一种非常便利的策略是使用测试驱动开发的一些（而不是全部）概念，同时开发代码以进行测试。这种策略有几个优点：

- 确保一旦软件的一部分已经通过测试，其他部分的更改就不会对已经开发的部分产生负面影响。测试是逐步进行的。所有测试都总是通过的，评估新功能和先前存在的代码的有效性，从而加快了开发速度。
- 如果来自其他开发人员的贡献已经添加到软件包中，则修订任务将大大简化。在你的存储库中激活持续集成（Continous Integration，CI）系统可确保每个贡献都必须正确编译并通过所有测试（功能和样式）。通过这种方式，评审人员将着重验证代码是否正确执行其工作。
- 许多质量保证程序要求对软件进行测试。说"等我完成后再做测试"是一种谬论：你不会去做它们，或者这将是一项乏味的过程，它不会帮助你，所以它们很可能是不完整和无效的。

ROS2 提供了许多测试工具，我们可以轻松使用这些工具。让我们从单元测试开始。ROS2 使用 GoogleTest 来测试 C++ 代码。为了在包中使用测试，需要在 package.xml 中包含一些包：

```xml
package.xml

<test_depend>ament_lint_auto</test_depend>
<test_depend>ament_lint_common</test_depend>
<test_depend>ament_cmake_gtest</test_depend>
```

<test_depend> 标记只包含测试包所需的那些依赖项。可以编译一个工作空间，它仅编译包，而不包括测试，因此这些包不会在依赖项中考虑：

```
$ colcon build --symlink-install --packages-select br2_vff_avoidance
  --cmake-args -DBUILD_TESTING=off
```

如包结构所示，有一个包含测试的 C++ 文件（vff_test.cpp）的 tests 目录。要编译它，这些句子应该在 CMakeLists.txt 中：

```cmake
CMakeLists.txt

if(BUILD_TESTING)
  find_package(ament_lint_auto REQUIRED)
  ament_lint_auto_find_test_dependencies()

  set(ament_cmake_cpplint_FOUND TRUE)
  ament_lint_auto_find_test_dependencies()

  find_package(ament_cmake_gtest REQUIRED)
  add_subdirectory(tests)
endif()
```

```cmake
tests/CMakeLists.txt

ament_add_gtest(vff_test vff_test.cpp)
ament_target_dependencies(vff_test ${dependencies})
target_link_libraries(vff_test ${PROJECT_NAME})
```

一旦在包中引入了测试基础设施，就可以了解如何进行单元测试。在开发 AvoidanceNode::get_vff 方法时，可以检查它是否正确工作。只需要创建几个合成的 sensor_msgs::msg::LaserScan 消息，然后检查此函数在所有情况下返回正确的值。在此文件中，开发了 8 种不同的情况。让我们来看看其中的一些情况：

```cpp
tests/vff_test.cpp

sensor_msgs::msg::LaserScan get_scan_test_1(rclcpp::Time ts)
{
  sensor_msgs::msg::LaserScan ret;
  ret.header.stamp = ts;
```

```
tests/vff_test.cpp

    ret.angle_min = -M_PI;
    ret.angle_max = M_PI;
    ret.angle_increment = 2.0 * M_PI / 16.0;
    ret.ranges = std::vector<float>(16, std::numeric_limits<float>::infinity());

    return ret;
}

sensor_msgs::msg::LaserScan get_scan_test_5(rclcpp::Time ts)
{
    sensor_msgs::msg::LaserScan ret;
    ret.header.stamp = ts;
    ret.angle_min = -M_PI;
    ret.angle_max = M_PI;
    ret.angle_increment = 2.0 * M_PI / 16.0;
    ret.ranges = std::vector<float>(16, 5.0);
    ret.ranges[10] = 0.3;

    return ret;
}
```

每个函数返回一个 sensor_msgs::msg::LaserScan 消息，就好像它是由具有 16 个不同值的激光器产生的，这些值有规律地分布在 [–π, π] 范围内。在 get_scan_test_1 中，它模拟在任何情况下都没有检测到障碍物的情况。在 get_scan_test_5 中，它模拟在位置 10 有一个障碍物，其响应角度为 $-\pi + 10 \times \frac{2\pi}{16} = 0.785$。

为了访问要测试的方法，因为它不是公开的，所以可以方便地将其设置为 protected 并实现一个类来访问这些函数：

```
tests/vff_test.cpp

class AvoidanceNodeTest : public br2_vff_avoidance::AvoidanceNode
{
public:
    br2_vff_avoidance::VFFVectors
    get_vff_test(const sensor_msgs::msg::LaserScan & scan)
    {
        return get_vff(scan);
    }

    visualization_msgs::msg::MarkerArray
    get_debug_vff_test(const br2_vff_avoidance::VFFVectors & vff_vectors)
    {
        return get_debug_vff(vff_vectors);
    }
};
```

可以将所有需要的测试都放在同一个文件中。它们都是用 TEST(id, sub_id) 宏来定义的，并且在其中编写一个目标是测试代码功能的程序，就像编写函数一样。以下是针对 get_vff 的单元测试：

```
tests/vff_test.cpp

TEST(vff_tests, get_vff)
{
```

```cpp
tests/vff_test.cpp

  auto node_avoidance = AvoidanceNodeTest();
  rclcpp::Time ts = node_avoidance.now();

  auto res1 = node_avoidance.get_vff_test(get_scan_test_1(ts));
  ASSERT_EQ(res1.attractive, std::vector<float>({1.0f, 0.0f}));
  ASSERT_EQ(res1.repulsive, std::vector<float>({0.0f, 0.0f}));
  ASSERT_EQ(res1.result, std::vector<float>({1.0f, 0.0f}));

  auto res2 = node_avoidance.get_vff_test(get_scan_test_2(ts));
  ASSERT_EQ(res2.attractive, std::vector<float>({1.0f, 0.0f}));
  ASSERT_NEAR(res2.repulsive[0], 1.0f, 0.00001f);
  ASSERT_NEAR(res2.repulsive[1], 0.0f, 0.00001f);
  ASSERT_NEAR(res2.result[0], 2.0f, 0.00001f);
  ASSERT_NEAR(res2.result[1], 0.0f, 0.00001f);

  auto res5 = node_avoidance.get_vff_test(get_scan_test_5(ts));
  ASSERT_EQ(res5.attractive, std::vector<float>({1.0f, 0.0f}));
  ASSERT_LT(res5.repulsive[0], 0.0f);
  ASSERT_LT(res5.repulsive[1], 0.0f);
  ASSERT_GT(atan2(res5.repulsive[1], res5.repulsive[0]), -M_PI);
  ASSERT_LT(atan2(res5.repulsive[1], res5.repulsive[0]), -M_PI_2);
  ASSERT_LT(atan2(res5.result[1], res5.result[0]), 0.0);
  ASSERT_GT(atan2(res5.result[1], res5.result[0]), -M_PI_2);
}

int main(int argc, char ** argv)
{
  rclcpp::init(argc, argv);
  testing::InitGoogleTest(&argc, argv);
  return RUN_ALL_TESTS();
}
```

ASSERT_* 宏根据输入检查预期值。ASSERT_EQ 验证两个值是否相等。当比较浮点数时，最好使用 ASSERT_NEAR，它检查两个值是否相等，其第三个参数指定了特定的范围。ASSERT_LT 验证第一个值是否"小于"第二个值。ASSERT_GT 验证第一个值是否"大于"第二个值，依此类推。

例如，情况 5（角度为 0.785 的障碍物）验证了排斥向量的坐标是否为负值，它的角度是否在 [-π, -π/2] 范围内（它是一个与角度 0.785 相反的向量），并且结果向量是否在 [0, -π/2] 范围内。如果这些都是真的，那么算法是正确的。对每个读数的预期值进行这些检查，并注意极端和意外的情况，比如测试 1。

也可以进行集成测试。由于节点是对象，因此需要实例化它们并模拟它们的操作。例如，在接收测试消息时测试 AvoidanceNode 发布的速度。让我们看看如何做：

```cpp
tests/vff_test.cpp

TEST(vff_tests, ouput_vels)
{
  auto node_avoidance = std::make_shared<AvoidanceNodeTest>();

  // Create a testing node with a scan publisher and a speed subscriber
  auto test_node = rclcpp::Node::make_shared("test_node");
  auto scan_pub = test_node->create_publisher<sensor_msgs::msg::LaserScan>(
    "input_scan", 100);

  geometry_msgs::msg::Twist last_vel;
  auto vel_sub = test_node->create_subscription<geometry_msgs::msg::Twist>(
```

```cpp
tests/vff_test.cpp

    "output_vel", 1, [&last_vel] (geometry_msgs::msg::Twist::SharedPtr msg) {
      last_vel = *msg;
    });

  ASSERT_EQ(vel_sub->get_publisher_count(), 1);
  ASSERT_EQ(scan_pub->get_subscription_count(), 1);

  rclcpp::Rate rate(30);
  rclcpp::executors::SingleThreadedExecutor executor;
  executor.add_node(node_avoidance);
  executor.add_node(test_node);

  // Test for scan test #1
  auto start = node_avoidance->now();
  while (rclcpp::ok() && (node_avoidance->now() - start) < 1s) {
    scan_pub->publish(get_scan_test_1(node_avoidance->now()));
    executor.spin_some();
    rate.sleep();
  }
  ASSERT_NEAR(last_vel.linear.x, 0.3f, 0.0001f);
  ASSERT_NEAR(last_vel.angular.z, 0.0f, 0.0001f);

  // Test for scan test #2
}
```

- 创建一个 AvoidanceNodeTest（也可以是 AvoidanceNode）节点进行测试。
- 创建一个名为 test_node 的通用节点，它用于创建激光扫描发布者和速度订阅者。
- 在创建速度订阅者时，指定了一个 lambda 函数作为回调。这个 lambda 函数访问 last_vel 变量，以在 output_vel 主题接收到的最后一个消息更新它。
- 创建执行器并将两个节点添加到其中以执行它们。
- 在以 30Hz 的频率对 input_scan 进行第二次发布时，传感器读数对应合成读数。
- 最后，验证发布的速度是否正确。

要运行这些测试，可以通过在包的测试目录中运行二进制文件来实现，在构建目录中：

```
$ cd ~/bookros2_ws
$ build/br2_vff_avoidance/tests/vff_test

[==========] Running 2 tests from 1 test case.
[----------] Global test environment set-up.
[----------] 2 tests from vff_tests
[ RUN      ] vff_tests.get_vff
[       OK ] vff_tests.get_vff (18 ms)
[ RUN      ] vff_tests.ouput_vels
[       OK ] vff_tests.ouput_vels (10152 ms)
[----------] 2 tests from vff_tests (10170 ms total)

[----------] Global test environment tear-down
[==========] 2 tests from 1 test case ran. (10170 ms total)
[  PASSED  ] 2 tests.
```

要运行此包的所有测试，包括样式测试，请使用 colcon：

```
$ colcon test --packages-select br2_vff_avoidance
```

如果测试失败，那么请去目录 log/latest_test/br2_vff_avoidance/stdout_stderr.log 查看哪些失败了。在文件末尾，有一个失败测试的总结。例如，此消息最后表明测试 3、4、5 和 7 失败（为此解释特意添加了错误）：

```
log/latest_test/br2_vff_avoidance/stdout_stderr.log

56% tests passed, tests failed out of 9

Label Time Summary:
copyright     =   0.37 sec*proc (1 test)
cppcheck      =   0.44 sec*proc (1 test)
cpplint       =   0.45 sec*proc (1 test)
flake8        =   0.53 sec*proc (1 test)
gtest         =  10.22 sec*proc (1 test)
lint_cmake    =   0.34 sec*proc (1 test)
linter        =   3.88 sec*proc (8 tests)
pep257        =   0.38 sec*proc (1 test)
uncrustify    =   0.38 sec*proc (1 test)
xmllint       =   0.99 sec*proc (1 test)

Total Test time (real) =  14.11 sec

The following tests FAILED:
      [ 3 - cpplint (Failed)]
      [ 4 - flake8 (Failed)]
      [ 5 - lint_cmake (Failed)]
      [ 7 - uncrustify (Failed)]
Errors while running CTest
```

该文件中的每一行都以与测试相对应的节号开头。例如，前往第 3、4 和 7 节看看这些错误：

```
log/latest_test/br2_vff_avoidance/stdout_stderr.log

3: br2_vff_avoidance/tests/vff_test.cpp:215:  Add #include <memory> for
   make_shared<>.  [build/include_what_you_use] [4]
3: br2_vff_avoidance/include/br2_vff_avoidance/AvoidanceNode.hpp:15:  #ifndef header
   guard has wrong style, please use: BR2_VFF_AVOIDANCE__AVOIDANCENODE_HPP_
   [build/header_guard] [5]

4: ./launch/avoidance_vff.launch.py:34:3: E111 indentation is not a multiple of four
4:   ld.add_action(vff_avoidance_cmd)
4:

7: --- src/br2_vff_avoidance/AvoidanceNode.cpp
7: +++ src/br2_vff_avoidance/AvoidanceNode.cpp.uncrustify
7: @@ -100,2 +100 @@
7: -  if (distance_min < OBSTACLE_DISTANCE)
7: -  {
7: +  if (distance_min < OBSTACLE_DISTANCE) {
7: @@ -109 +108 @@
7: -    vff_vector.repulsive[0] = cos(oposite_angle)*complementary_dist;
7: +    vff_vector.repulsive[0] = cos(oposite_angle) * complementary_dist;
7:
7: Code style divergence in file 'tests/vff_test.cpp':
```

- 第 3 节中的错误对应于 cpplint，这是一个 C++ 的静态代码检查工具。第一个错误表明必须添加一个头文件，因为有一些函数是在这个头文件中声明的。第二个错误表明 AvoidanceNode.hpp 中的头文件保护风格是不正确的，并指出了应该使用哪种风格。

- 第 4 节中的错误对应 flake8，这是一个 Python 的静态代码检查工具。这个错误表明启动器文件使用了错误的缩进，它应该是 4 的倍数的空格。
- 标记为 7 的错误对应 uncrustify，这是另一个 C++ 的静态代码检查工具。类似于 diff 命令输出的格式，它告诉所写的代码和应该遵循良好风格的代码之间的差异。在这种情况下，它指出 AvoidanceNode.cpp 的第 100 行的 if 块的开始应该在 if 的同一行。第二个错误表明运算符两侧应该有空格。

当你第一次面对编码风格问题时，可能会感觉这是一项令人望而却步的任务。你会想知道为什么它指出的样式比你的更好。的确，你已经使用该风格多年了，你对你的源代码的外观感到非常自豪。你不明白为什么在 C++ 中必须使用两个空格进行缩进，而不能使用制表符，或者为什么在 while 打开块时要在同一行打开，而你总是在下一行打开。

第一个原因是，它指示了良好的风格样式。例如，cpplint 使用的是 Google C++ 样式指南⊖，这是一个被大多数软件开发公司采用的广泛接受的风格样式指南。

第二个原因是，如果你想为一个 ROS2 项目或存储库做出贡献，则必须遵循这个风格。几乎所有接受贡献的代码库都是用连续集成系统来通过这些测试。想象一下，你是维护一个项目的负责人，你希望所有的代码都有一致的风格。如果每次拉取请求都需要讨论风格问题，那将是一场噩梦。我与同事最糟糕的一次讨论是关于使用制表符还是空格的问题。这是一个没有解决方案的讨论。使用一个标准可以解决这些问题。

此外，最后一个原因是这会让你成为一个更好的程序员。大多数的风格规则都有实际的原因。随着时间的推移，你将自然地在编写代码时应用这些风格，从而使你的代码在编写过程中就有良好的风格。

5.2 跟踪对象

这部分内容分析了一个包含其他反应式行为的项目。在这种情况下，机器人的头部行为会跟踪与特定颜色匹配的对象。在这个项目中引入了几个新概念：

- **图像分析**：目前为止，我们使用了一个相对简单的传感器。图像提供了更复杂的感知信息，可以从中提取很多信息。请记住，人工智能中有一个关键领域是计算机视觉，它是机器人的主要传感器之一。我们将展示如何使用

⊖ https://google.github.io/styleguide/cppguide.html。

OpenCV 参考库来处理这些图像。
- **关节级控制**：在之前的项目中，控制命令主要涉及向机器人发送速度信号。在这种情况下，我们将学习如何直接向机器人的颈部关节发送位置命令。
- **生命周期节点**：ROS2 提供了一种特殊类型的节点，即生命周期节点。这种节点对于控制节点的生命周期非常有用，包括启动、激活和停用。

5.2.1 感知和执行模型

本项目使用机器人相机的图像作为信息来源。每当一个节点在 ROS2 中传输一个非压缩图像时，它都使用相同类型的消息：sensor_msgs/msg/Image。所有在 ROS2 中支持的相机的驱动程序都使用它。查看信息格式：

```
$ ros2 interface show sensor_msgs/msg/Image
# This message contains an uncompressed image
# This message contains an uncompressed

std_msgs/Header header  # Header timestamp should be acquisition time of image
                        # Header frame_id should be optical frame of camera
                        # origin of frame should be optical center of camera
                        # +x should point to the right in the image
                        # +y should point down in the image
                        # +z should point into to plane of the image
                        # If the frame_id and the frame_id of the CameraInfo
                        # message associated with the image conflict
                        # the behavior is undefined

uint32 height           # image height, that is, number of rows
uint32 width            # image width, that is, number of columns

# The legal values for encoding are in file src/image_encodings.cpp
# If you want to standardize a new string format, join
# ros-users@lists.ros.org and send an email proposing a new encoding.

string encoding         # Encoding of pixels -- channel meaning, ordering, size
                        # from the list in include/sensor_msgs/image_encodings.hpp

uint8 is_bigendian      # is this data bigendian?
uint32 step             # Full row length in bytes
uint8[] data            # actual matrix data, size is (step * rows)
```

摄像机驱动程序通常会发布（仅一次，以临时本地 QoS）有关摄像机参数的信息，并将其作为一个 sensor_msgs/msg/CameraInfo 消息，包括内参和畸变参数、投影矩阵等。有了这些信息，我们可以处理立体图像，例如，我们可以将这些信息与深度图像结合起来，以重建 3D 场景。校准摄像机⊖的过程涉及计算在此消息中发布

⊖ http://wiki.ros.org/image_pipeline。

的值。虽然本章未使用此消息格式，但阅读此消息格式是一个很好的练习。

虽然可以使用简单的 sensor_msgs/msg/Image 发布者或订阅者，但在使用不同的传输策略（压缩、流编解码器等）来处理图像时，通常会使用特定的发布者或订阅者。开发人员使用它们但并不需要了解图像是如何传输的——他们只看到一个 sensor_msgs/msg/Image。检查可用的传输插件，输入：

```
$ ros2 run image_transport list_transports
Declared transports:
image_transport/compressed
image_transport/compressedDepth
image_transport/raw
image_transport/theora

Details:
----------
...
```

运行模拟 Tiago，并检查主题，可以看到 2D 图像有多个主题：

```
$ ros2 topic list
/head_front_camera/image_raw/compressed
/head_front_camera/image_raw/compressedDepth
/head_front_camera/image_raw/theora
/head_front_camera/rgb/camera_info
/head_front_camera/rgb/image_raw
```

开发人员并没有逐一地创建所有主题，而是使用了一个 image_transport::Publisher 来生成所有主题，并考虑到可用的传输插件。同样，为了获得图像，使用 image_transport::Subscriber 是很方便的。如果图像很大或者网络可靠性不是最好的，那么使用压缩图像可能是好的。代价是源和目标上的 CPU 负载要多一点。

图像消息格式是用于传输图像的，而不是用于处理图像的。直接将图像作为原始字节序列是不常见的。通常的方法是使用一些图像处理库，其中使用最广泛的是 OpenCV[⊖]。OpenCV 提供了数百种计算机视觉算法。

OpenCV 用于处理图像的主要数据类型是 cv::Mat。ROS2 提供了将 sensor_msgs/msg/Image 转换为 cv::Mat 的工具，反之亦然。

```
void image_callback(const sensor_msgs::msg::Image::ConstSharedPtr & msg)
{
  cv_bridge::CvImagePtr cv_ptr;
  cv_ptr = cv_bridge::toCvCopy(msg, sensor_msgs::image_encodings::BGR8);
  cv::Mat & image_src = cv_ptr->image;

  sensor_msgs::msg::Image image_out = *cv_ptr->toImageMsg();
}
```

⊖ https://docs.opencv.org/5.x/d1/dfb/intro.html。

在我们项目的感知模型中，图像的分割是通过颜色设置来完成的。在 HSV 中工作很方便，而不是 RGB，RGB 我们接收消息的编码。HSV 编码表示一个像素的颜色有三个组成部分：色相、饱和度和值。在 HSV 中工作使我们能够更牢固地建立光照变化的颜色范围，因为这是 V 分量主要负责的内容，如果范围更宽，即使光照变化，我们也可以继续检测到相同的颜色。

下面的代码将 cv::mat 转换为 HSV，并计算出一个图像掩码，其像素与 Gazebo 中 Tiago 的默认模拟世界中的家具颜色相匹配。图 5.4 展示了使用 HSV 范围过滤器进行颜色的对象检测。

```
cv::Mat img_hsv;
cv::cvtColor(cv_ptr->image, img_hsv, cv::COLOR_BGR2HSV);

cv::Mat1b filtered;
cv::inRange(img_hsv, cv::Scalar(15, 50, 20), cv::Scalar(20, 200, 200), filtered);
```

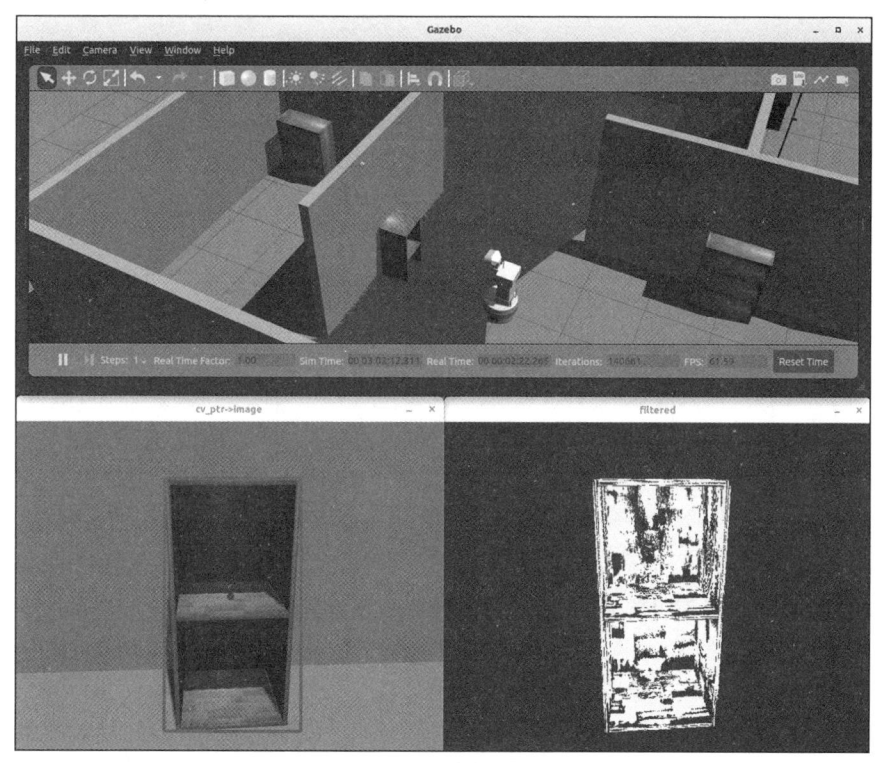

图 5.4 使用 HSV 范围过滤器进行颜色的对象检测（见彩插）

在这个项目中，图像处理的最终输出是一个类型为 vision_msgs/msg/Detection2D 的消息（请自行检查此消息中的字段），我们使用它的 header、bbox 和 source_img

字段。不需要使用所有字段。包含原始图像是为了拥有进行检测时图像的尺寸，其重要性将在下文中展示。

动作模型是机器人头部的位置控制。机器人有两个控制摄像机指向位置的关节：用于水平控制（平移）的 head_1_joint 和用于垂直控制（倾斜）的 head_2_joint。

在 ROS2 中，关节的控制是通过 ros2_control[一]框架完成的。模拟 Tiago 机器人的开发者使用了一个轨迹控制器（joint_trajectory_controller）来控制机器人颈部的两个关节。通过两个主题（头控制器主题如图 5.5 所示），可以读取关节的状态，并以一组要在特定时间点到达的路径点的消息格式（见图 5.6）发送命令。路径点包括位置，还可以包括速度、加速度和力矩，以及应用的起始时间。

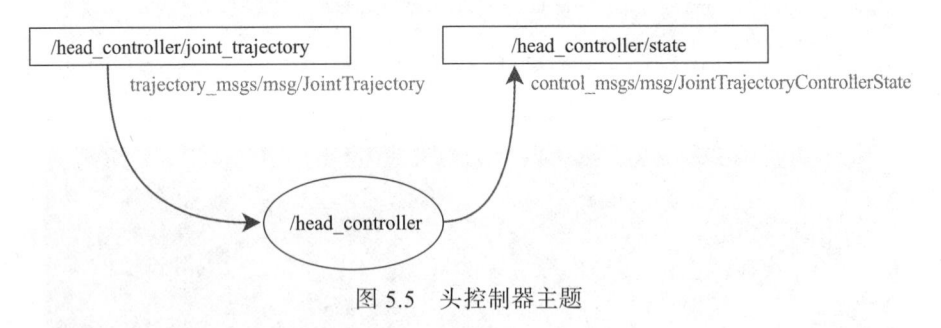

图 5.5　头控制器主题

图 5.6　trajectory_msgs/msg/JointTrajectory 消息格式

获取物体相对于机器人的 3D 位置并计算颈部关节的位置以使其在图像中居中，这可能是在真实的主动视觉系统中的一个合适的解决方案，但对目前来说相当复杂。我们将简单地在图像域中实现一个控制。

控制机器人颈部的节点接收两个值（称为 error），它们分别表示平移和倾斜的当前位置和期望位置之间的差异。如果某个值为 0，则表示它在期望的位置；如果它小于 0，

[一] http://control.ros.org/index.html。
[二] 消息中的一个字段，包含元数据、时间戳和坐标系信息。——译者注
[三] 表示轨迹中的一系列路径点。——译者注

则关节必须向一个方向运动；如果大于 0，则关节需要向另一个方向移动。每个关节的取值范围为 [−1, +1]，平移/倾斜控制图如图 5.7 所示。由于该节点执行迭代控制，颈部运动可以非常快，因此 PID 控制器将控制每个关节的位置以被命令纠正其速度。

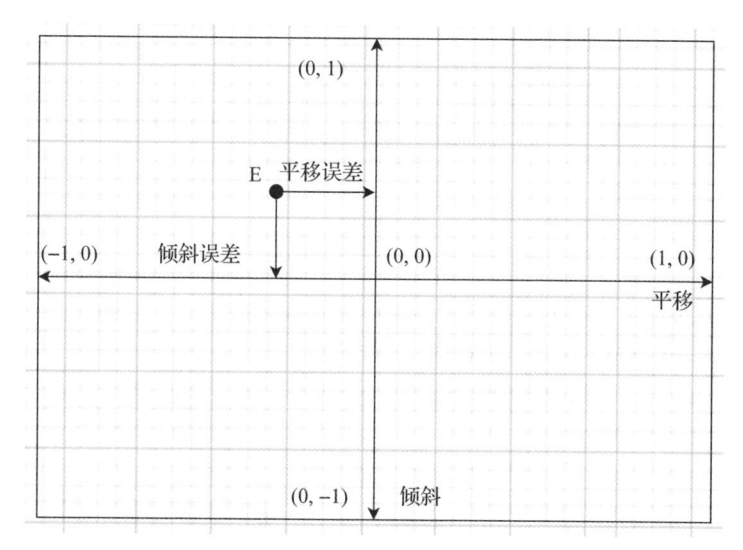

图 5.7 平移/倾斜控制图。E 表示期望位置，error_* 表示当前位置与期望的平移/倾斜位置之间的差异

注：error_pan——平移误差，指的是平移控制中的错误或异常状态
　　error_tilt——倾斜误差，指的是倾斜控制中的错误或异常状态
　　pan——平移，指的是摄像头的水平移动，从一侧到另一侧
　　tilt——倾斜，指的是摄像头的垂直移动，向上或向下倾斜

5.2.2　计算图

对象跟踪项目的计算图（见图 5.8）显示了如何将此问题划分为同一进程中的三个节点。之所以这样做，是因为每个节点（ObjectDetector 和 HeadController）可以单独执行，并在其他问题中被重用（我们将在下一章中这样做）。每个节点都以可重用的方式设计，输入和输出是通用的，而不是与这个问题强耦合的。

在这个计算图中，HeadController 与其他节点的表示方式不同。这个节点将被实现为生命周期节点，我们将在 5.2.3 节中解释。现在，我们说它就像一个标准节点，但它可以在其操作期间被激活和停用。

HeadController 接收一个平移/倾斜速度，每个速度值的范围为 [−1, 1]。请注意，由于没有符合我们问题的标准 ROS2 消息（我们可以使用 geometry_msgs/msg/

Pose2D，忽略字段 theta），因此我们创建了一个包含所需信息的自定义 br2_tracking_msgs/msg/PanTiltCommand 消息。下面将看到我们是如何创建自定义消息的。

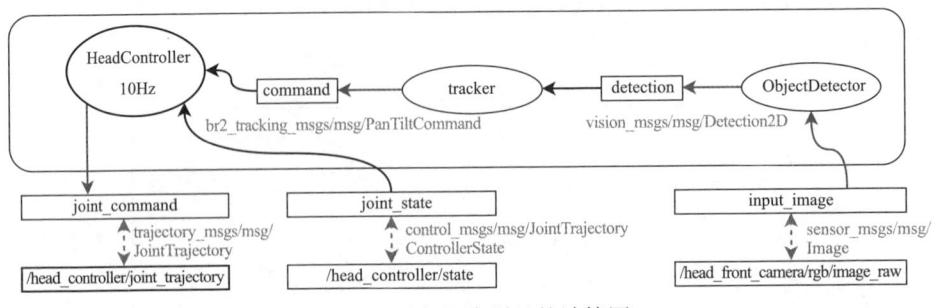

```
┌─────────────────────────────────────────────────────────────────────┐
│   ┌──────────────┐                                                  │
│   │ HeadController│                                                 │
│   │    10Hz      │ ← command ← tracker ← detection ← ObjectDetector │
│   └──────────────┘                                                  │
│              br2_tracking_msgs/msg/PanTiltCommand                   │
│                                            vision_msgs/msg/Detection2D│
└─────────────────────────────────────────────────────────────────────┘
     ↓                         ↑↓                          ↑
┌──────────────┐         ┌──────────────┐          ┌──────────────┐
│joint_command │         │ joint_state  │          │ input_image  │
└──────────────┘         └──────────────┘          └──────────────┘
  trajectory_msgs/msg/    control_msgs/msg/JointTrajectory  sensor_msgs/msg/
  JointTrajectory         ControllerState                   Image
 /head_controller/joint_trajectory  /head_controller/state  /head_front_camera/rgb/image_raw
```

图 5.8 对象跟踪项目的计算图

ObjectDetector 为每个图像发布图像中家具的检测结果。它将返回检测的坐标（以像素为单位）以及对象的边界框。

ObjectDetector 的输出与 HeadController 的输入不完全匹配。ObjectDetector 以像素为单位发布其输出。在本例中，图像分辨率为 640×480，因此水平 X 分量的范围为 [0, 640]，垂直 Y 分量的范围为 [0, 480]。因此，我们创建了一个 tracker 节点，它有一个简单的任务，就是将 ObjectDetector 的输出调整为 HeadController 的输入，在图像中创建一个控件，移动头部，以便检测到的对象始终位于图像的中心。

5.2.3 生命周期节点

到目前为止，我们已经看到 ROS2 中的节点是 Node 类的对象，它们继承方法，使我们能够与其他节点通信或获取信息。在 ROS2 中，有一种节点被称为 Life-CycleNode，其生命周期使用状态和它们之间的转换来定义：

- 当一个生命周期节点（LifeCycleNode）被创建时，它处于未配置（Unconfigured）状态，并且必须触发配置（configure）过渡才能进入非激活（Inactive）状态。
- 生命周期节点在处于激活（Active）状态时工作，它可以通过激活（activate）过渡从非激活状态转变过来，也可以通过停用（deactivate）过渡从激活状态转回非激活状态。
- 每个过渡都可以执行必要的任务和检查。如果一个过渡没有满足其过渡代码中指定的条件，则可能会失败且不转变。
- 在出现错误的情况下，节点可以进入终止（Finalized）状态。
- 当节点完成其任务时，它可以过渡到终止状态。

请参见图 5.9 中这些生命周期节点的状态和过渡的示意图。生命周期节点提供了一种节点执行模型，允许

- 使它们可预测。例如，在 ROS2 中，参数应该只在配置过渡中读取。
- 当有多个节点时，我们可以协调它们的启动。我们可以定义特定的节点在配置完毕之前不被激活。我们还可以在启动中指定一些顺序。
- 在编程上，它提供了除了构造函数之外的另一种启动其组件的选项。在 C++ 中，直到其构造函数完成，一个节点才完全建立。如果我们需要一个对当前节点的 shared_ptr 共享指针，那么这通常会带来问题。

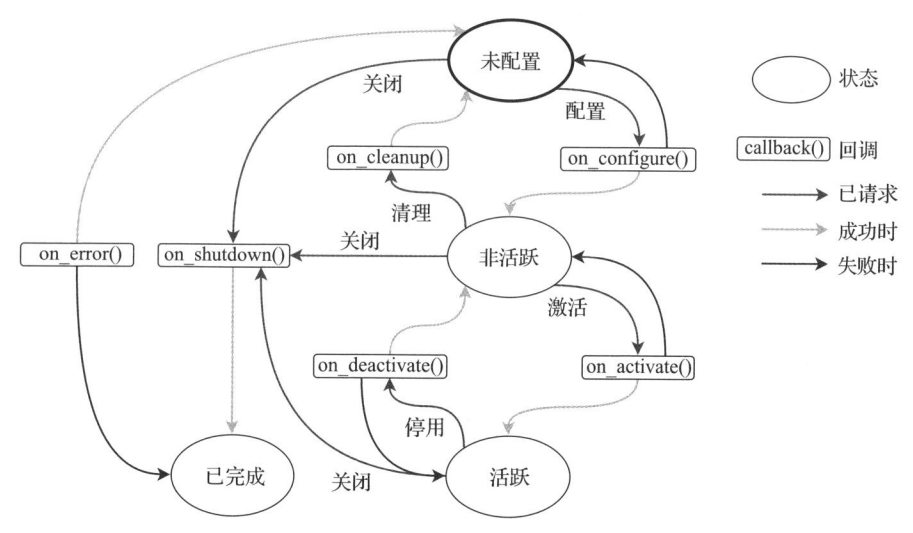

图 5.9 生命周期节点的状态和过渡图（见彩插）

表 5.1 展示了生命周期节点的状态和过渡图的术语及说明。

表 5.1 生命周期节点的状态和过渡图的术语及说明

术语	说明
状态	状态机中的一个基本单元，表示系统在某一时刻的情况或模式
回调	在某些事件发生时自动执行的函数或方法
已请求	转换或操作已被请求，但尚未执行的状态
成功时	在转换成功完成后调用的回调函数
失败时	在操作或转换失败时调用的回调
未配置	系统或组件尚未完成初始化配置的状态
关闭	系统或组件正在关闭或已关闭的状态
配置	系统或组件正在进行配置的过程

㊀ 该对照表为译者注释，原书无此部分。——编辑注

(续)

术语	说明
清理	系统或组件正在执行清理操作的状态
非活跃	系统或组件当前处于非活跃或等待状态
激活	使系统或组件从非活跃状态变为活跃状态的操作
停用	使系统或组件从活跃状态变为非活跃状态的操作
活跃	系统或组件当前处于正常运行的活跃状态
已完成	系统或组件已完成其操作,进入最终状态
on_cleanup()	清理时,在系统或组件清理时调用的回调函数
on_configure()	配置时,在系统或组件配置时调用的回调函数
on_error()	错误时,在系统或组件发生错误时调用的回调函数
on_shutdown()	关闭时,在系统或组件关闭时调用的回调函数
on_deactivate()	停用时,在系统或组件停用时调用的回调函数
on_activate()	激活时,在系统或组件激活时调用的回调函数

一个例子可能是一个传感器驱动程序。如果无法访问物理设备,则它无法转换为非活动状态。此外,所有设备的初始设置时间都将设置为此状态,以即时激活。另一个例子是机器人驱动程序的启动。除非其所有传感器/执行器节点都处于活动状态,否则它将无法启动。

5.2.4 创建自定义消息

我们先前指定了 HeadController 节点的输入类型为 br2_tracking_msgs/msg/PanTiltCommand,因为没有符合我们需要的消息类型。ROS2 的一个黄金法则是,如果已经有标准实现,则不需要创建自定义的消息,因为我们可以从此消息的可用工具中受益。在这种情况下,没有标准可以满足我们的目的。此外,这是创建自定义消息的完美借口。

首先,在创建新消息(通常是新接口)时,即使在特定包的上下文中,也强烈建议你制作一个以 _msgs 结尾的单独包。未来可能存在需要接收此新类型消息的工具,但我们并不一定要依赖创建的包。

接下来,我们展示了包含消息 br2_tracking_msgs/msg/PanTiltCommand 定义的 br2_tracking_msgs 包的结构:

```
Package br2_tracking_msgs

br2_tracking_msgs/
├── CMakeLists.txt
├── msg
│   └── PanTiltCommand.msg
└── package.xml
```

包含接口的软件包除有一个 package.xml 和一个 CMakeLists.txt 文件外，还有一个用于定义每种接口类型（消息、服务或动作）的目录。在我们的示例中，它是一条消息，因此，我们将有一个 msg 目录，其中包含用于定义每个新消息的 .msg 文件。请参阅 PanTiltCommand 消息的定义：

```
msg/PanTiltCommand.msg
```
```
float64 pan
float64 tilt
```

CMakeLists.txt 中的重要部分是 rosidl_generate_interfaces 语句，我们在其中指定接口定义的位置：

```
CMakeLists.txt
```
```
find_package(ament_cmake REQUIRED)
find_package(builtin_interfaces REQUIRED)
find_package(rosidl_default_generators REQUIRED)

rosidl_generate_interfaces(${PROJECT_NAME}
  "msg/PanTiltCommand.msg"
  DEPENDENCIES builtin_interfaces
)

ament_export_dependencies(rosidl_default_runtime)
ament_package()
```

5.2.5 跟踪实现

br2_tracking 软件包的结构如下所示，遵循上面已经推荐的指导原则。

```
Package br2_vff_avoidance
```
```
br2_tracking
├── CMakeLists.txt
├── config
│   └── detector.yaml
├── include
│   └── br2_tracking
│       ├── HeadController.hpp
│       ├── ObjectDetector.hpp
│       └── PIDController.hpp
├── launch
│   └── tracking.launch.py
├── package.xml
├── src
│   ├── br2_tracking
│   │   ├── HeadController.cpp
│   │   ├── ObjectDetector.cpp
│   │   └── PIDController.cpp
│   └── object_tracker_main.cpp
└── tests
    ├── CMakeLists.txt
    └── pid_test.cpp
```

- HeadController 和 ObjectDetector 节点将作为库被单独编译，与 object_tracker_main.cpp 主程序无关。后者将位于 src 中，而节点的头文件将位于 include/

- 库还包括一个用于 PID 控制器的类，它用于 HeadController。
- 启动器将使用必要的参数和重映射启动可执行文件。
- 存在一个包含 YAML 文件的配置目录，其中包含 ObjectDetector，它将用于检测 Gazebo 中 Tiago 默认阶段的家具的 HSV 范围。
- 测试目录包括 PID 控制器的测试。

读者会注意到在这个结构中没有跟踪器节点的文件。这个节点非常简单，已经在 object_tracker_main.cpp 中实现。代码如下：

```cpp
// src/object_tracker_main.cpp
auto node_detector = std::make_shared<br2_tracking::ObjectDetector>();
auto node_head_controller = std::make_shared<br2_tracking::HeadController>();
auto node_tracker = rclcpp::Node::make_shared("tracker");

auto command_pub = node_tracker->create_publisher<br2_tracking_msgs::msg::PanTiltCommand>(
  "/command", 100);
auto detection_sub = node_tracker->create_subscription<vision_msgs::msg::Detection2D>(
  "/detection", rclcpp::SensorDataQoS(),
  [command_pub](vision_msgs::msg::Detection2D::SharedPtr msg) {
    br2_tracking_msgs::msg::PanTiltCommand command;
    command.pan = (msg->bbox.center.x / msg->source_img.width) * 2.0 - 1.0;
    command.tilt = (msg->bbox.center.y / msg->source_img.height) * 2.0 - 1.0;
    command_pub->publish(command);
  });

rclcpp::executors::SingleThreadedExecutor executor;
executor.add_node(node_detector);
executor.add_node(node_head_controller->get_node_base_interface());
executor.add_node(node_tracker);
```

node_tracker 是一个通用的 ROS2 节点，我们在这个节点上构建了一个发布到 /command 主题的发布者和一个发布到 /detection 的订阅者。我们已经将订阅者回调指定为 lambda 函数，该函数从输入消息中获取检测对象的像素位置以及图像的大小，并生成用于 HeadController 节点的输入，这遵循了图 5.7 中已经展示的方案。

请注意，在将 node_head_controller 节点添加到执行器时，我们使用了 get_node_base_interface 方法。这是因为它是一个生命周期节点（LifeCycleNode），正如我们之前介绍的，而 add_node 目前还不支持直接添加这种类型的节点。幸运的是，我们可以通过被生命周期节点（LifeCycleNode）和常规节点支持的基础接口来实现。

ObjectDetector 将是一个 rclcpp::Node，具有使用图像传输的图像订阅者和 2D 检测消息发布者。检测过程中将使用两个成员变量。

```cpp
// include/br2_tracking/ObjectDetector.hpp
class ObjectDetector : public rclcpp::Node
{
public:
  ObjectDetector();

  void image_callback(const sensor_msgs::msg::Image::ConstSharedPtr & msg);
```

```
include/br2_tracking/ObjectDetector.hpp
```
```
private:
  image_transport::Subscriber image_sub_;
  rclcpp::Publisher<vision_msgs::msg::Detection2D>::SharedPtr detection_pub_;

  // HSV ranges for detection [h - H] [s - S] [v - V]
  std::vector<double> hsv_filter_ranges_ {0, 180, 0, 255, 0, 255};
  bool debug_ {true};
};
```

这些变量都具有默认值，将使用参数进行初始化。它们是 HSV 颜色范围和一个变量，该变量在默认情况下会导致带有检测结果的窗口，它用于调试目的。

```
src/br2_tracking/ObjectDetector.cpp
```
```
ObjectDetector::ObjectDetector()
: Node("object_detector")
{
  declare_parameter("hsv_ranges", hsv_filter_ranges_);
  declare_parameter("debug", debug_);

  get_parameter("hsv_ranges", hsv_filter_ranges_);
  get_parameter("debug", debug_);
}
```

在执行程序时，将在配置目录中指定参数文件以设置颜色滤镜。

```
config/detector.yaml
```
```
/object_detector:
  ros__parameters:
    debug: true
    hsv_ranges:
      - 15.0
      - 20.0
      - 50.0
      - 200.0
      - 20.0
      - 200.0
```

这个节点旨在为每个到达的图像获得一个结果，因此只要有订阅者订阅这个结果，处理就直接在回调中完成。

创建一个 image_transport::Subscriber 与创建一个 rclcpp::Subscription 非常相似。因为第一个参数是一个 rclcpp::Node*，所以我们使用 this。第四个参数指示传输方法，在这种情况下是 raw。我们在最后的参数中调整服务质量，以适应传感器中常见的设置。

```
src/br2_tracking/ObjectDetector.cpp
```
```
ObjectDetector::ObjectDetector()
: Node("object_detector")
{
  image_sub_ = image_transport::create_subscription(
```

src/br2_tracking/ObjectDetector.cpp
```
    this, "input_image", std::bind(&ObjectDetector::image_callback, this, _1),
    "raw", rclcpp::SensorDataQoS().get_rmw_qos_profile());

  detection_pub_ = create_publisher<vision_msgs::msg::Detection2D>("detection", 100);
}

void
ObjectDetector::image_callback(const sensor_msgs::msg::Image::ConstSharedPtr & msg)
{
  if (detection_pub_->get_subscription_count() == 0) {return;}
  ...
  vision_msgs::msg::Detection2D detection_msg;
  ...
  detection_pub_->publish(detection_msg);
}
```

图像处理已经在前面的章节中介绍过了。一旦图像消息被转换成 cv::Mat，我们就继续将其从 RGB 转换为 HSV，并进行颜色过滤。cv::boundingRect 函数从颜色过滤得到的掩码中计算一个边界框。cv::moments 函数计算这些像素的质心。

src/br2_tracking/ObjectDetector.cpp
```
const float & h = hsv_filter_ranges_[0];
const float & H = hsv_filter_ranges_[1];
const float & s = hsv_filter_ranges_[2];
const float & S = hsv_filter_ranges_[3];
const float & v = hsv_filter_ranges_[4];
const float & V = hsv_filter_ranges_[5];

cv_bridge::CvImagePtr cv_ptr;
try {
  cv_ptr = cv_bridge::toCvCopy(msg, sensor_msgs::image_encodings::BGR8);
} catch (cv_bridge::Exception & e) {
  RCLCPP_ERROR(get_logger(), "cv_bridge exception: %s", e.what());
  return;
}

cv::Mat img_hsv;
cv::cvtColor(cv_ptr->image, img_hsv, cv::COLOR_BGR2HSV);

cv::Mat1b filtered;
cv::inRange(img_hsv, cv::Scalar(h, s, v), cv::Scalar(H, S, V), filtered);

auto moment = cv::moments(filtered, true);
cv::Rect bbx = cv::boundingRect(filtered);

auto m = cv::moments(filtered, true);

if (m.m00 < 0.000001) {return;}

int cx = m.m10 / m.m00;
int cy = m.m01 / m.m00;

vision_msgs::msg::Detection2D detection_msg;
detection_msg.header = msg->header;
detection_msg.bbox.size_x = bbx.width;
detection_msg.bbox.size_y = bbx.height;
detection_msg.bbox.center.x = cx;
detection_msg.bbox.center.y = cy;
detection_msg.source_img = *cv_ptr->toImageMsg();
detection_pub_->publish(detection_msg);
```

在之前的代码中，处理了图像，获取了过滤像素在 filtered 中的边界框 bbx，并且连同质心（cx，cy）一起发布。此外，还填充了可选的 source_img 字段，因为我们在 object_tracker_main.cpp 中需要图像的大小。

HeadController 的实现稍微复杂一些。首先它是一个生命周期节点，以及它的控制循环只在节点处于活跃状态时被调用。让我们看看节点的声明，仅仅关注其控制基础设施的部分：

```
include/br2_tracking/HeadController.hpp

class HeadController : public rclcpp_lifecycle::LifecycleNode
{
public:
  HeadController();

  CallbackReturn on_configure(const rclcpp_lifecycle::State & previous_state);
  CallbackReturn on_activate(const rclcpp_lifecycle::State & previous_state);
  CallbackReturn on_deactivate(const rclcpp_lifecycle::State & previous_state);

  void control_sycle();

private:
  rclcpp_lifecycle::LifecyclePublisher<trajectory_msgs::msg::JointTrajectory>::SharedPtr
    joint_pub_;
  rclcpp::TimerBase::SharedPtr timer_;
};
```

LifecycleNode::create_subscription 方法返回的是一个 rclcpp_lifecycle::LifecyclePublisher 而不是一个 rclcpp::Publisher。尽管它们的功能相似，但需要激活它才能使用。

生命周期节点可以在派生类中重新定义在状态转换时触发的函数。这些函数可以返回 SUCCESS 或 FAILURE。如果返回 SUCCESS，则允许进行状态转换。如果返回 FAILURE，则不转换到新状态。在基类中，所有方法都返回 SUCCESS，但开发者可以重新定义它们以建立拒绝条件。

在这种情况下，导致进入非活动状态的转换（在配置时）和在活动和非活动之间的转换（在激活和非激活时）被重新定义：

```
src/br2_tracking/HeadController.cpp

HeadController::HeadController()
: LifecycleNode("head_tracker")
{
  joint_pub_ = create_publisher<trajectory_msgs::msg::JointTrajectory>(
    "joint_command", 100);
}

CallbackReturn
HeadController::on_configure(const rclcpp_lifecycle::State & previous_state)
{
  return CallbackReturn::SUCCESS;
}

CallbackReturn
HeadController::on_activate(const rclcpp_lifecycle::State & previous_state)
{
  joint_pub_->on_activate();
  timer_ = create_wall_timer(100ms, std::bind(&HeadController::control_sycle, this));
```

```
src/br2_tracking/HeadController.cpp
  return CallbackReturn::SUCCESS;
}

CallbackReturn
HeadController::on_deactivate(const rclcpp_lifecycle::State & previous_state)
{
  joint_pub_->on_deactivate();
  timer_ = nullptr;

  return CallbackReturn::SUCCESS;
}

void
HeadController::control_sycle()
{
}
```

所有前面的转换都返回 SUCCESS，因此所有转换都已完成。在开发激光驱动器时，如果设备未被发现或无法访问，则某些过渡（配置或激活）将失败。

上述代码有两个有趣的方面需要解释：
- contrcol_cycle 方法包含我们的控制逻辑，并设置为以 10Hz 的频率运行。请注意，定时器是在 on_activate 时创建的，这是节点转换到激活状态的时候。同样，禁用这个定时器只需要通过转换到非激活状态。这样，contrcol_cycle 就不会被调用，控制逻辑只会在节点处于激活状态时执行。
- 发布者必须在 on_activate 中被激活，在 on_deactivate 中被停用。

HeadController 节点将会迭代执行，通过 /joint_state 主题接收颈部关节的当前状态，通过 /command 主题接收移动命令。在这种模式中，last_state_ 和 last_command_ 中的值都会被存储起来，以便在执行控制逻辑的下一个周期时使用。同时，也会保存最后接收到的命令的时间戳。当停止接收命令时，机器人应该返回到初始位置。

```
include/br2_tracking/HeadController.hpp
class HeadController : public rclcpp_lifecycle::LifecycleNode
{
public:
  void joint_state_callback(
    control_msgs::msg::JointTrajectoryControllerState::UniquePtr msg);
  void command_callback(br2_tracking_msgs::msg::PanTiltCommand::UniquePtr msg);

private:
  rclcpp::Subscription<br2_tracking_msgs::msg::PanTiltCommand>::SharedPtr command_sub_;
  rclcpp::Subscription<control_msgs::msg::JointTrajectoryControllerState>::SharedPtr
    joint_sub_;
  rclcpp_lifecycle::LifecyclePublisher<trajectory_msgs::msg::JointTrajectory>::SharedPtr
    joint_pub_;

  control_msgs::msg::JointTrajectoryControllerState::UniquePtr last_state_;
  br2_tracking_msgs::msg::PanTiltCommand::UniquePtr last_command_;
  rclcpp::Time last_command_ts_;
};
```

```
src/br2_tracking/HeadController.cpp
void
HeadController::joint_state_callback(
  control_msgs::msg::JointTrajectoryControllerState::UniquePtr msg)
{
  last_state_ = std::move(msg);
}

void
HeadController::command_callback(br2_tracking_msgs::msg::PanTiltCommand::UniquePtr msg)
{
  last_command_ = std::move(msg);
  last_command_ts_ = now();
}
```

control_msgs::msg::JointTrajectoryControllerState 的格式旨在报告受控关节的名称，以及期望的、当前的和错误的轨迹：

```
$ ros2 interface show control_msgs/msg/JointTrajectoryControllerState

std_msgs/Header header
string[] joint_names
trajectory_msgs/JointTrajectoryPoint desired
trajectory_msgs/JointTrajectoryPoint actual
trajectory_msgs/JointTrajectoryPoint error # Redundant, but useful
```

一开始，使用 trajectory_msgs::msg::JointTrajectory 可能看起来很复杂，但如果我们分析以下代码，就会发现其实并不复杂，这段代码是一个命令，它用于将机器人的颈部置于初始状态，同时查看图 5.6：

```
src/br2_tracking/HeadController.cpp
CallbackReturn
HeadController::on_deactivate(const rclcpp_lifecycle::State & previous_state)
{
  trajectory_msgs::msg::JointTrajectory command_msg;
  command_msg.header.stamp = now();
  command_msg.joint_names = last_state_->joint_names;
  command_msg.points.resize(1);
  command_msg.points[0].positions.resize(2);
  command_msg.points[0].velocities.resize(2);
  command_msg.points[0].accelerations.resize(2);
  command_msg.points[0].positions[0] = 0.0;
  command_msg.points[0].positions[1] = 0.0;
  command_msg.points[0].velocities[0] = 0.1;
  command_msg.points[0].velocities[1] = 0.1;
  command_msg.points[0].accelerations[0] = 0.1;
  command_msg.points[0].accelerations[1] = 0.1;
  command_msg.points[0].time_from_start = rclcpp::Duration(1s);

  joint_pub_->publish(command_msg);

  return CallbackReturn::SUCCESS;
}
```

- 关节名称字段是一个 std::vector<std::string>，它包含被控制的关节的名称。在这种情况下有两个，它们与已经在状态消息中的相同。
- 将发送一个单一的路径点（因此将点字段调整为 1），其中必须为每个关节指定位置、速度和加速度（因为有两个关节，所以这些字段的每个字段都调整为两个）。位置 0 对应于关节名称为 0 的关节，以此类推。
- time_from_start 表示到达命令位置所需的时间。由于它是在停用之前发送的最后一个命令（这就是为什么它的期望位置是 0 的原因），因此需要一秒钟的时间才能不强制颈部电机。

颈关节的控制器是通过发送包含位置的命令来控制的，但从 ObjectDetector 接收到的是应该执行的速度控制，以便将检测到的物体置于图像中心。

第一种实现方法可能是将当前位置与接收到的控制命令结合起来作为位置发送：

```
src/br2_tracking/HeadController.cpp
command_msg.points[0].positions[0] = last_state_->actual.positions[0] - last_command_->pan;
command_msg.points[0].positions[1] = last_state_->actual.positions[1] -last_command_->tilt;
```

如果读者使用这种实现方式，他会发现，即使 ObjectDetector 与 HeadDetector 的频率差异很小，但如果我们想要保持足够的反应速度，那么机器人的头部可能会开始振荡，试图将在图像中检测到的物体置于中心。机器人很难对检测到的物体保持稳定聚焦。这个问题在工程中通过 PID 控制器来解决，每个关节一个，它在限制速度的同时也吸收颈部的小的不希望的振荡。

```
include/br2_tracking/HeadController.hpp
class HeadController : public rclcpp_lifecycle::LifecycleNode
{
private:
  PIDController pan_pid_, tilt_pid_;
};
```

对于每个 PID，定义比例组件 K_p、积分组件 K_i 和微分组件 K_d 的值。在此不深入细节，因为本书的目标不是深入描述背后的控制理论。直观地说，比例组件使我们更接近目标。积分组件补偿持续的偏差，这些偏差使我们远离目标。微分组件试图在接近控制目标时减少微小变化。图 5.10 是这个 PID 的示意图。

来自跟踪器的控制命令是 PID 应该尝试保持为 0 的值，因此它是时刻 t 的误差 $e(t)$。PID 的每个组件都被分别计算，然后相加以获得要应用的控制 $u(t)$。发送到关节的位置将是关节的当前位置加上 $u(t)$。该系统是反馈的，因为在时间 $t + 1$ 时，控制的效果会反映在物体在图像中心的位置变化上。

我们的 PID 从指定四个值开始：PID 中预期的最小和最大输入参考值以及产生的最小和最大输出值。负输入会产生负输出：

```
src/br2_tracking/PIDController.hpp
class PIDController
{
```

```
src/br2_tracking/PIDController.hpp
```
```cpp
public:
  PIDController(double min_ref, double max_ref, double min_output, double max_output);

  void set_pid(double n_KP, double n_KI, double n_KD);
  double get_output(double new_reference);
};
```

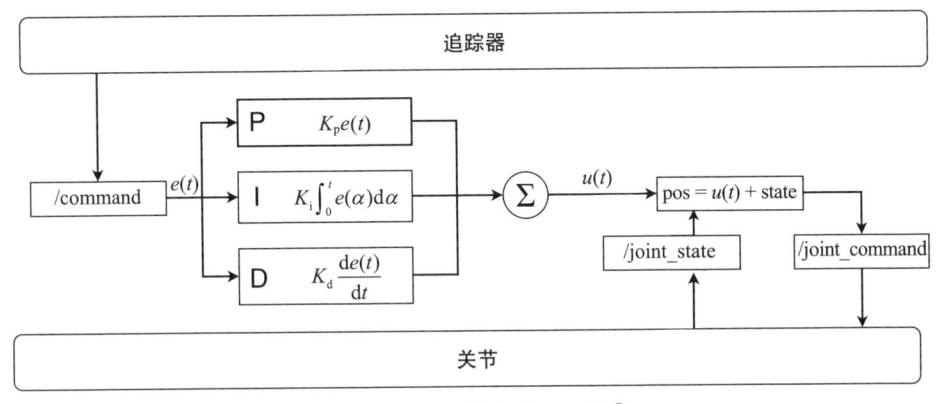

```
                    ┌──────────────── 追踪器 ────────────────┐
                    │   ┌─────────────────────┐              │
                    │   │  P    K_p e(t)      │              │
                    │   ├─────────────────────┤              │
         e(t)       │   │  I    K_i ∫₀ᵗ e(α)dα │──→ Σ ─u(t)→ pos = u(t) + state ──┐
    /command ──────→┤   ├─────────────────────┤              │                    │
                    │   │  D    K_d de(t)/dt  │              │   /joint_state    /joint_command
                    │   └─────────────────────┘              │       ↑                │
                    └────────────────────────────────────────┘       │                │
                                           ↑                         │                │
                    ┌──────────────────── 关节 ───────────────────────┴────────────────┘
```

图 5.10 一个关节的 PID 图[○]

```
src/br2_tracking/HeadController.cpp
```
```cpp
HeadController::HeadController()
: LifecycleNode("head_tracker"),
  pan_pid_(0.0, 1.0, 0.0, 0.3),
  tilt_pid_(0.0, 1.0, 0.0, 0.3)
{
}
CallbackReturn
HeadController::on_configure(const rclcpp_lifecycle::State & previous_state)
{
  pan_pid_.set_pid(0.4, 0.05, 0.55);
  tilt_pid_.set_pid(0.4, 0.05, 0.55);
}
void
HeadController::control_sycle()
{
  double control_pan = pan_pid_.get_output(last_command_->pan);
  double control_tilt = tilt_pid_.get_output(last_command_->tilt);

  command_msg.points[0].positions[0] = last_state_->actual.positions[0] - control_pan;
  command_msg.points[0].positions[1] = last_state_->actual.positions[1] - control_tilt;
}
```

[○] 图中的追踪器为用于追踪或监控某个变量或状态的系统组件；关节为机器人或机械系统中可以移动的部分；P 为比例，比例控制是控制理论中的一种基本控制方式，它与误差成比例；I 为积分，积分控制关注误差随时间的累积效果，它通过消除稳态误差来提供系统的稳态响应；D 为微分，微分控制预测误差的趋势，即误差变化的快慢，它可以增加系统的稳定性和响应速度。——译者注

5.2.6 执行跟踪器

在 object_tracker_main.cpp 主程序中，所有节点都被创建并添加到一个执行器中。在开始旋转节点之前，我们触发了 node_head_controller 节点的配置（configure）转换。当请求时，节点将准备好被激活。

```
src/object_tracker_main.cpp
rclcpp::executors::SingleThreadedExecutor executor;
executor.add_node(node_detector);
executor.add_node(node_head_controller->get_node_base_interface());
executor.add_node(node_tracker);

node_head_controller->trigger_transition(
  lifecycle_msgs::msg::Transition::TRANSITION_CONFIGURE);
```

启动器重映射主题，并使用 HSV 过滤器参数加载文件：

```
launch/tracking.launch.py
params_file = os.path.join(
  get_package_share_directory('br2_tracking'),
  'config',
  'detector.yaml'
  )
object_tracker_cmd = Node(
  package='br2_tracking',
  executable='object_tracker',
  parameters=[{
    'use_sim_time': True
  }, params_file],
  remappings=[
    ('input_image', '/head_front_camera/rgb/image_raw'),
    ('joint_state', '/head_controller/state'),
    ('joint_command', '/head_controller/joint_trajectory')
  ],
  output='screen'
)
```

启动 Tiago 模拟（默认为家庭环境）gazebo：

```
$ ros2 launch br2_tiago sim.launch.py
```

在另一个终端，启动项目：

```
$ ros2 launch br2_tracking tracking.launch.py
```

检测窗口不会显示，直到第一个对象被检测到，但是 HeadController 处于非激活状态，并且不会进行跟踪。

看看我们如何在运行时管理 LifeCycleNode，比如 head_tracker（HeadController 节点的名称）。让我们的项目继续运行，使机器人跟踪一个物体，如图 5.11 所示。

使用以下命令，检查当前正在运行的 LifeCycle 节点：

```
$ ros2 lifecycle nodes

/head_tracker
```

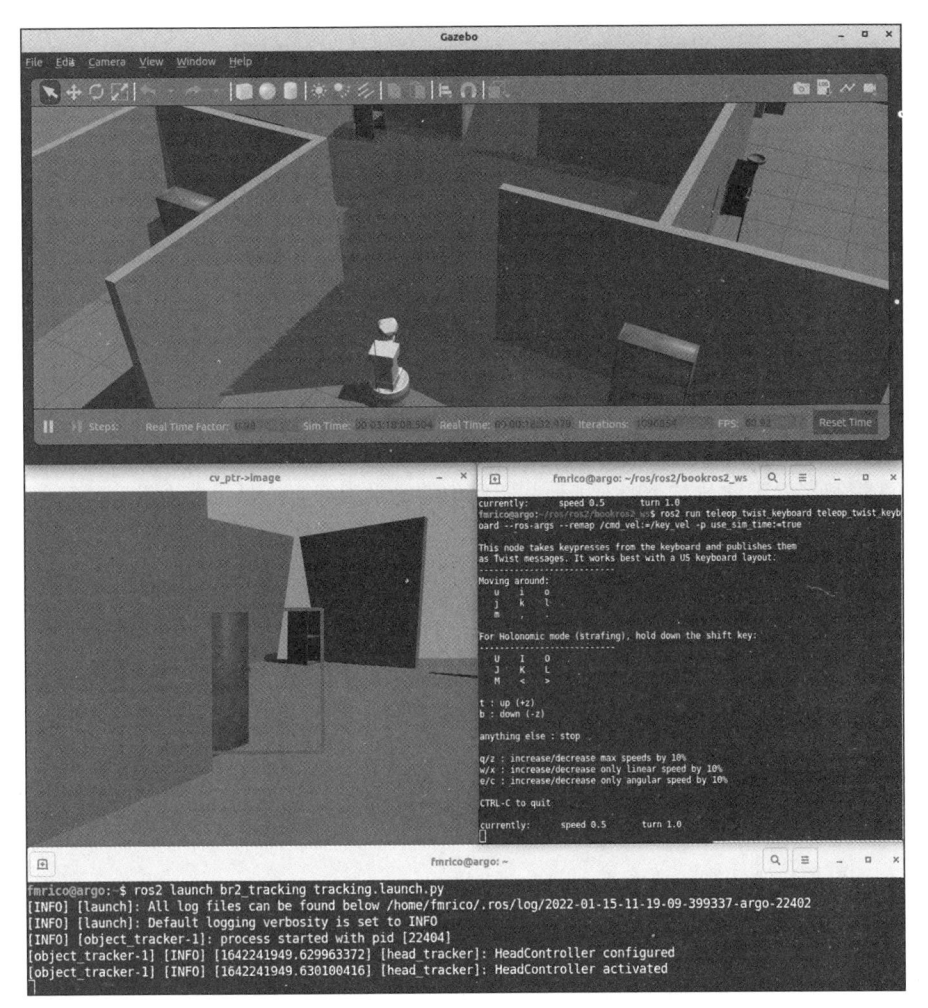

图 5.11 运行中的跟踪项目（见彩插）

现在验证它当前所处的状态：

```
$ ros2 lifecycle get /head_tracker
inactive [3]
```

很好。正如预期的那样，LifeCycleNode 处于非激活状态。获取可以从当前状态触发的转换：

```
$ ros2 lifecycle list /head_tracker
- cleanup [2]
```

```
                    Start: inactive
                    Goal: cleaningup
- activate [3]
                    Start: inactive
                    Goal: activating
- shutdown [6]
                    Start: inactive
                    Goal: shuttingdown
```

激活节点开始跟踪检测到的对象:

```
$ ros2 lifecycle set /head_tracker activate
Transitioning successful
```

在第三个终端运行远程操作器,使机器人走向家具。然后,机器人将移动(当头部控制器处于激活状态时)头部到图像中的家具中心。一旦机器人没有感知到物体,它就会将头部移动到初始位置:

```
$ ros2 run teleop_twist_keyboard teleop_twist_keyboard --ros-args -r
cmd_vel:=key_vel
```

停用节点并检查机器人的颈部如何返回到其初始位置。请记住,这是在此转换中的命令的操作,在 on_deactivate 方法中。

```
$ ros2 lifecycle set /head_tracker deactivate
Transitioning successful
```

若要再次激活它,请输入:

```
$ ros2 lifecycle set /head_tracker activate
Transitioning successful
```

建议的练习

1)在 AvoidanceNode 中,不是使用最近的障碍物,而是使用所有检测到的附近障碍物来计算排斥向量。

2)在 ObjectDetector 中,不是计算一个包含所有通过滤波器的像素的边界框,而是为每个独立的对象计算一个边界框。发布最近检测到的对象对应的边界框。

3)尝试使 HeadController 反应更快。

CHAPTER 6

第 6 章

用行为树对机器人行为进行编程

近年来,行为树是一种非常流行的用于控制机器人的方法[4],它应用广泛,主要用于电子游戏和机器人的领域。人们通常会将行为树与有限状态机进行比较,但实际上它们是不同的近似方法。在使用有限状态机(FSM)开发机器人行为时,我们考虑状态和转换。当使用行为树时,我们会考虑序列、回退以及许多流资源,这些流资源赋予了它们极强的表现力。在本章中,我们将用一个具体的示例来实现使用有限状态机完成的"碰撞与前进"任务,并看看这两种方法的差异。

6.1 行为树

行为树(Behavior Tree,BT)是一种数学模型,用于编码系统的控制。行为树可实现结构化自主智能体(如机器人或计算机游戏中的虚拟实体)不同任务之间的切换。它是一种分层数据结构,即从根节点开始递归定义,有若干子节点。每个子节点又可以有更多的子节点,依此类推。没有子节点的节点通常称为树的叶子。

节点的基本操作是执行(tick)。当一个节点被执行时,它可以返回三个不同的值:

- 成功(SUCCESS):节点已成功完成任务。
- 失败(FAILURE):节点执行任务失败。
- 运行中(RUNNING):节点尚未完成任务。

行为树(BT)有四种不同类型的节点:

- 控制节点:这种类型的节点有 1–N 个子节点,其功能是将一次执行(tick)传递到它们的子节点。

- **装饰节点**：这种类型的节点是一种特殊的控制节点，装饰节点仅有一个子节点。
- **动作节点**：这种类型的节点是树的叶子节点，用户必须实现这些节点的行为，因为它们必须生成应用程序所需的控制。
- **条件节点**：这种类型的节点是一种特殊的动作节点，它无法返回运行中状态。在这种情况下，如果满足编码的条件，则返回成功；如果不满足，则返回失败。

图 6.1 展示了一个简单的行为树。当一个行为树被执行时，根节点就会被执行，直到其完成整个过程，也就是说，直到它返回成功或失败状态。

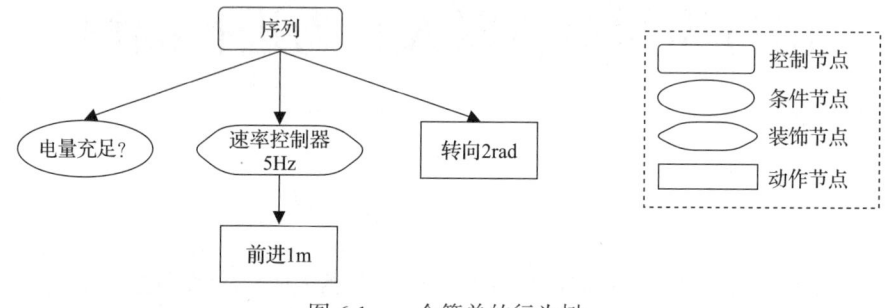

图 6.1　一个简单的行为树

- 根节点是一个类型为序列的控制节点。该节点按从左到右的顺序执行其子节点。当一个子节点返回成功时，序列节点会执行下一个子节点。如果子节点返回其他值，则序列节点将返回该值。
- 第一个子节点用于判断机器人是否有足够的电量，是一个条件节点。如果它返回成功，则表示机器人有足够的电池电量来执行任务，这样，序列节点就可以继续执行下一个子节点。如果它返回失败，则任务将被中止，因为执行行为树的结果将是失败。
- 前进动作节点命令机器人前进。在每次触发执行时，只要它前进未达到 1m，该节点就会返回运行中状态。当前进了指定的距离后，它将返回成功状态。
- 前进动作节点的父节点是一个装饰节点，用于控制其子节点的执行频率不大于 5Hz。同时，每次执行都返回子节点上一次执行返回的值。
- 转向动作节点与前进类似，但是要将机器人旋转 2rad。

可以使用用户创建的节点来扩展可用节点库。正如我们之前所说，用户必须实现动作节点，但如果我们需要其他类型的节点而当前可用库节点不满足的话，则我们也可以实现它。在上面的示例中，RateController 装饰节点不是行为树核心库的一部分，但可以由用户实现。

行为树控制行动决策的流程。叶子节点不用于实现复杂的算法或子系统。行为树叶子节点应协调机器人中的其他子系统。在 ROS2 中，这是通过发布或订阅主题、使用 ROS2 服务 / 动作来完成的，如图 6.2 所示。请注意，观察协调子系统的复杂性，而不是行为树叶子节点。

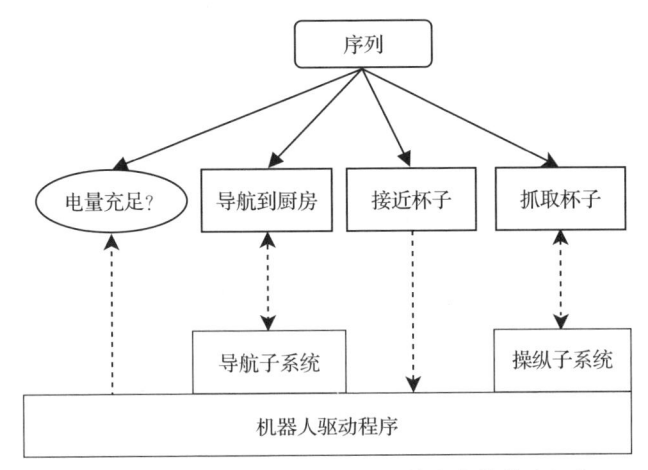

图 6.2 行为树的叶子节点通过发布 / 订阅（单向虚线箭头）或 ROS2 操作（双向虚线箭头）来控制机器人

第二个控制节点是回退节点。该节点可以使用回退策略，即当一个节点返回失败状态时应该做什么。

图 6.3 展示了具备回退策略的行为树。
- 回退节点执行第一个子节点。如果它返回失败状态，则执行下一个子节点。
- 如果第二个子节点返回成功状态，则回退节点返回成功状态。否则，它会执行下一个子节点。
- 如果所有子节点都返回失败状态，则回退节点返回失败状态。

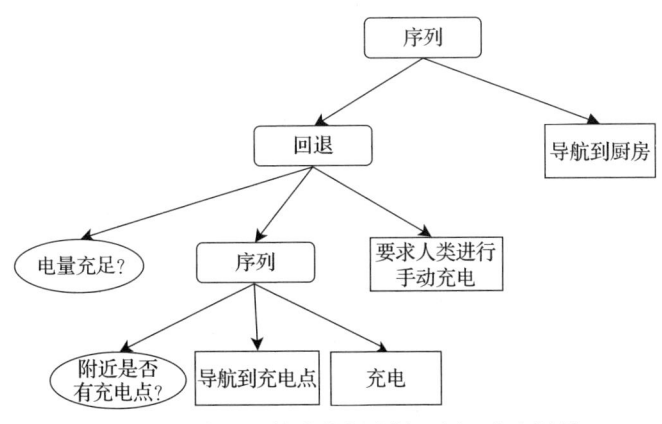

图 6.3 具备回退策略的行为树，用于充电场景

在行为树的开发周期中，我们可以识别以下两个阶段。
- **节点开发阶段**：动作节点和应用程序所需的其他节点在此阶段被设计、开发和编译。这些节点成为可用节点库的一部分，与行为树的核心节点类别相同。
- **部署阶段**：在此阶段，使用可用节点构成行为树。需要注意的是，可以使用相同的节点创建多个不同的行为树。如果节点设计得足够通用，那么在这个阶段，可以使用相同的节点定义机器人的不同行为。

在行为树中有一个所有节点都可以共享访问的键/值存储，我们称之为"黑板"（blackboard）。节点之间可以通过输入端口和输出端口交换信息。一个节点的输出端口通过黑板中的键连接到另一个节点的输入端口。虽然节点的端口（其类型和端口类）在编译时是需要确定的，但连接是在部署阶段建立的。

图 6.4 是使用黑板键连接端口的示例。DetectObject 动作节点负责检测某个对象，以便 InformHuman 节点通知机器人操作员。DetectObject 节点使用其输出端口 detected_id 将检测到的对象标识符发送到 InformHuman 节点的 object_id 端口。为此，它们使用了黑板的输入，其键 objID 的当前值为 cup。使用黑板的键不是强制性的。在部署阶段，该值可以是一个常量。

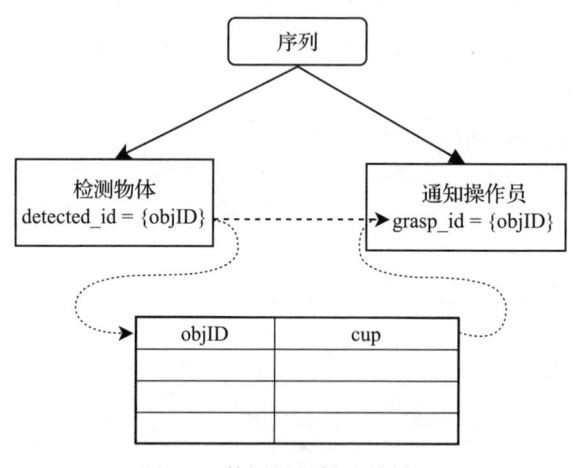

图 6.4 使用黑板键连接端口

行为树是用 XML 格式指定的。即便使用了编辑工具，如 Groot[一]，但它们生成的行为树也是 XML 格式的。如果将此行为树保存到磁盘并从应用程序中加载此文件，则对行为树的任何更改都不需要重新编译。该格式易于理解，并且广泛用于直接用 XML 设计行为树。以下代码显示了图 6.1 中行为树的两个等效的替代方案。

一 https://github.com/BehaviorTree/Groot。

Compact XML syntax

```xml
<BehaviorTree ID="BehaviorTree">
    <Sequence>
        <EnoughBattery/>
        <RateController Rate="5Hz">
            <GoForward distance="1.0"/>
        </RateController>
        <Turn angle="2.0"/>
    </Sequence>
</BehaviorTree>
```

Extended XML syntax

```xml
<?xml version="1.0"?>
<root main_tree_to_execute="BehaviorTree">

    <BehaviorTree ID="BehaviorTree">
        <Sequence>
            <Condition ID="EnoughBattery"/>
            <Decorator ID="RateController" Rate="5Hz">
                <Action ID="GoForward" distance="1.0"/>
            </Decorator>
            <Action ID="Turn" angle="2.0"/>
        </Sequence>
    </BehaviorTree>

    <TreeNodesModel>
        <Condition ID="EnoughBattery"/>
        <Action ID="GoForward">
            <input_port name="distance"/>
        </Action>
        <Decorator ID="RateController">
            <input_port name="Rate"/>
        </Decorator>
        <Action ID="Turn">
            <input_port name="angle"/>
        </Action>
    </TreeNodesModel>
</root>
```

表 6.1 总结了一些常用的控制节点的行为。该表显示了控制节点被执行时返回的内容（其值取决于被执行的子节点返回的内容）。对于序列和回退节点，表 6.1 还显示了如果再次执行此控制节点的操作：执行下一个、重新启动第一个子节点或坚持相同的子节点。

表 6.1 控制节点的行为总结。表格单元格颜色对于序列、回退和装饰节点进行了区分

控制节点类型	子节点返回的状态值		
	失败	成功	运行中
序列	返回失败并重新启动序列	执行下一个子节点，直到不存在下一个子节点，则返回成功	返回运行中并再次执行此节点
响应式序列	返回失败并重新启动序列	执行下一个子节点，直到不存在下一个子节点，则返回成功	返回运行中并重新启动序列
序列星	返回失败并重新执行此节点	执行下一个子节点，直到不存在下一个子节点，则返回成功	返回运行中并再次执行此节点

(续)

控制节点类型	子节点返回的状态值		
	失败	成功	运行中
回退	执行下一个节点，如果不存在下一个子节点，则返回失败	返回成功	返回运行中并再次执行此节点
响应式回退	执行下一个节点，如果不存在下一个子节点，则返回失败	返回成功	返回运行中并重新启动序列
反转节点	返回成功	返回失败	返回运行中
强制成功节点	返回成功	返回成功	返回运行中
强制失败节点	返回失败	返回失败	返回运行中
重复节点	返回失败	返回运行中，不断重复执行 N 次后才返回成功	返回运行中
重试节点	返回运行中，重试 N 次后仍然失败则返回失败	返回成功	返回运行中

让我们详细分析其中的一些控制节点：
- **序列节点**：在上一节中，我们使用了基本的序列节点。行为树允许具有不同行为的序列节点，这在某些应用中很有帮助。
    1) 序列（Sequence）：它会触发其第一个子节点。当它返回成功时状态，会触发下一个子节点，依此类推。如果任何子节点返回失败状态，则此节点将返回失败状态，并且如果再次被触发，则将从第一个子节点开始重新执行。图 6.5 展示了一个序列的示例，在该序列中，想要拍摄一张照片，就必须检查物体是否靠近以及相机是否就绪。一旦相机对准了主体，就可以拍照了。如果上述任何一个子节点失败，则整个序列都会失败。如果某个子节点已经成功完成，则不会重复执行其操作。

图 6.5 序列的示例

2) 响应式序列（ReactiveSequence）：通常在需要持续检查条件的情况下使用。如果任何子节点返回运行中状态，则序列将从头开始重新执行。这样，从第一个开始到上一次返回运行中状态的这些子节点都会被执行一遍。响应式序列的示例如图 6.6 所示。

3）序列星（SequenceStar）：它用于避免在某个子节点返回失败状态时重新启动序列。在下次执行时会直接对上次失败的子节点进行处理，而不是重新从序列的开头执行。序列星的示例如图 6.7 所示。

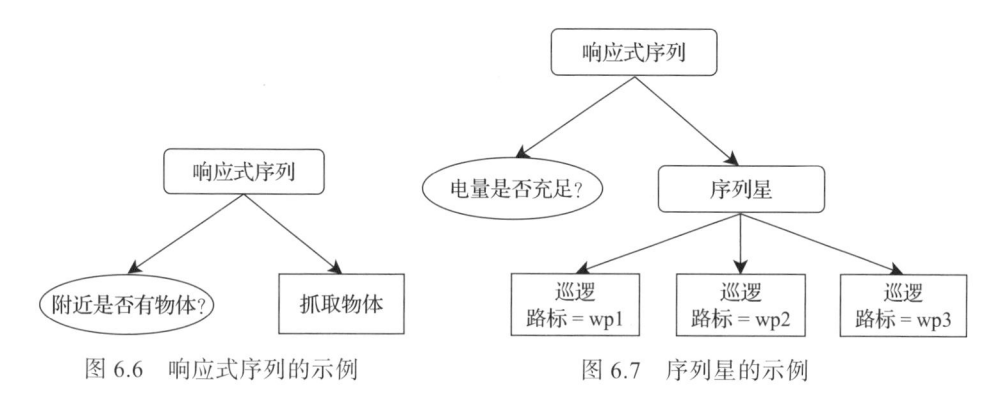

图 6.6 响应式序列的示例

图 6.7 序列星的示例

- 回退节点：它可以执行不同的策略以满足条件，直到找到一个成功的策略。
  1）回退（Fallback）：它是此控制节点的基本版本。子节点按顺序执行。当一个子节点返回失败状态时，它会继续执行下一个子节点。当一个子节点返回成功状态时，它会返回成功状态。
  2）响应式回退（ReactiveFallback）：它是回退的另一种版本。它与回退的区别在于，如果一个子节点返回运行中状态，那么序列将从头开始重新启动。下一个执行将再次从第一个子节点开始。这在第一个子节点是条件节点时非常有用，因为在执行后续操作时，必须检查条件是否满足。例如，图 6.8 是响应式回退的示例，在电池尚未充满的情况下，机器人正在执行充电的动作。

图 6.8 响应式回退的示例

- 装饰节点：它修改其唯一子节点的返回值。在重复节点（RepeatNode）和重试节点（RetryNode）的情况下，它通过其输入端口接收 N 次重复或重试。

6.2 使用行为树实现"碰撞与前进"任务

在本节中，我们将展示如何在 ROS2 包中实现动作节点，以及这些节点如何通过访问计算图来与其他节点进行通信。为此,我们将重新实现"碰撞与前进"示例,以了解存在的差异。

让我们从行为树的设计开始,一个完整的实现"碰撞与前进"任务的行为树如图 6.9 所示。很明显,我们需要涉及以下行为树节点(见图 6.10):

- 一个条件节点,根据从激光传感器接收到的信息指示是否存在障碍物(SUCCESS)或不存在障碍物(FAILURE)。
- 三个动作节点,使机器人转向、移动或前进,同时发布速度消息。后退(Back)和转向(Turn)节点在返回成功状态之前会保持运行中状态 3s。前进(Forward)将在所有执行周期中返回运行中状态。

图 6.9 一个完整的实现"碰撞与前进"任务的行为树

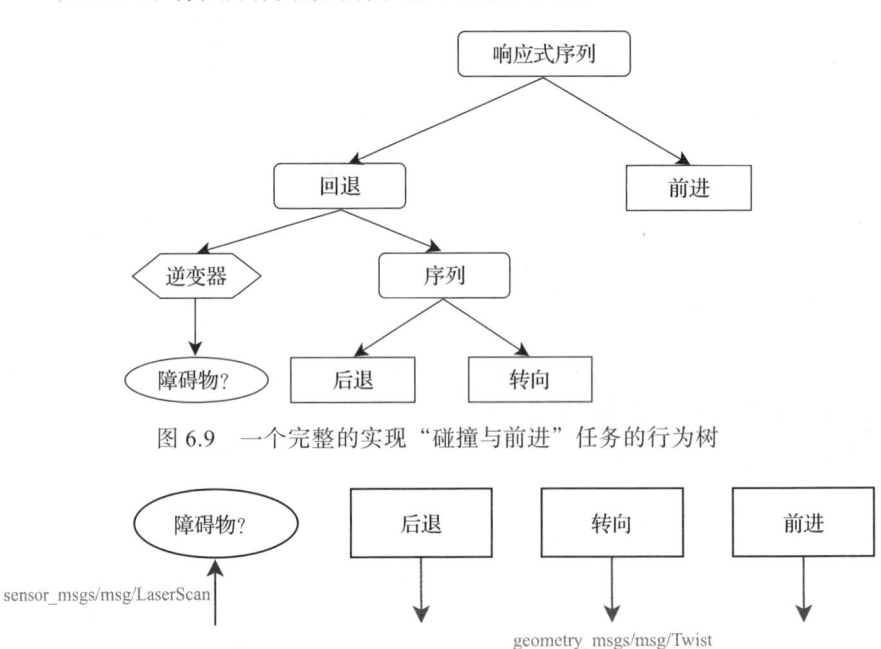

图 6.10 碰撞和前进中涉及的动作节点

计算图与图 3.4 类似,因此我们将跳过对其的解释。让我们专注于工作空间:

```
Package br2_bt_bumpgo
br2_bt_bumpgo
├── behavior_tree_xml
│   └── bumpgo.xml
├── cmake
│   └── FindZMQ.cmake
├── CMakeLists.txt
├── include
│   └── br2_bt_bumpgo
│       ├── Back.hpp
```

```
Package br2_bt_bumpgo
    │   ├── Forward.hpp
    │   ├── IsObstacle.hpp
    │   └── Turn.hpp
    ├── package.xml
    ├── src
    │   ├── br2_bt_bumpgo
    │   │   ├── Back.cpp
    │   │   ├── Forward.cpp
    │   │   ├── IsObstacle.cpp
    │   │   └── Turn.cpp
    │   └── bt_bumpgo_main.cpp
    └── tests
        ├── bt_action_test.cpp
        └── CMakeLists.txt
```

- 每个行为树节点都是一个 C++ 类。就像在实现 ROS2 节点时一样，我们在 src 目录中为源代码创建一个与包名称相同的目录，同时在 include 目录中为头文件创建一个相同名称的目录。
- 一个 tests 目录包含使用 gtest 进行测试的测试文件，以及一个用于手动测试行为树节点的程序，我们稍后会解释。
- 一个 cmake 目录包含一个 cmake 文件，它用于查找运行时需要的调试行为树的 ZMQ[⊖] 库。
- 一个 behavior_tree_xml 目录包含包中使用的行为树结构的 XML 文件。

6.2.1 使用 Groot 创建行为树

本节介绍一种用于开发和监控行为树的工具，即 Groot。该包中的行为树已经被创建好，但我们认为解释这个工具的工作原理是有帮助的。它对于监控运行时的性能非常有用，或者读者可能想要进行修改。

Groot 已包含在存储库依赖项中，因此要执行它，只需要输入：

```
$ ros2 run groot Groot
```

选择编辑器后，按照以下步骤操作：
- 将转向（Turn）节点、前进（Forward）节点、后退（Back）节点和是否障碍物（IsObstacle）节点添加到面板中。除了是否障碍物（IsObstacle）节点外，其余都是动作节点，因此，为是否障碍物节点添加一个输入端口，是否障碍物节点定义如图 6.11 所示。
- 保存面板。
- 根据图 6.12 生成行为树。
- 将行为树保存在 mr2_bt_bumpgo/behavior_tree_xml/bumpgo.xml。

[⊖] https://zeromq.org。

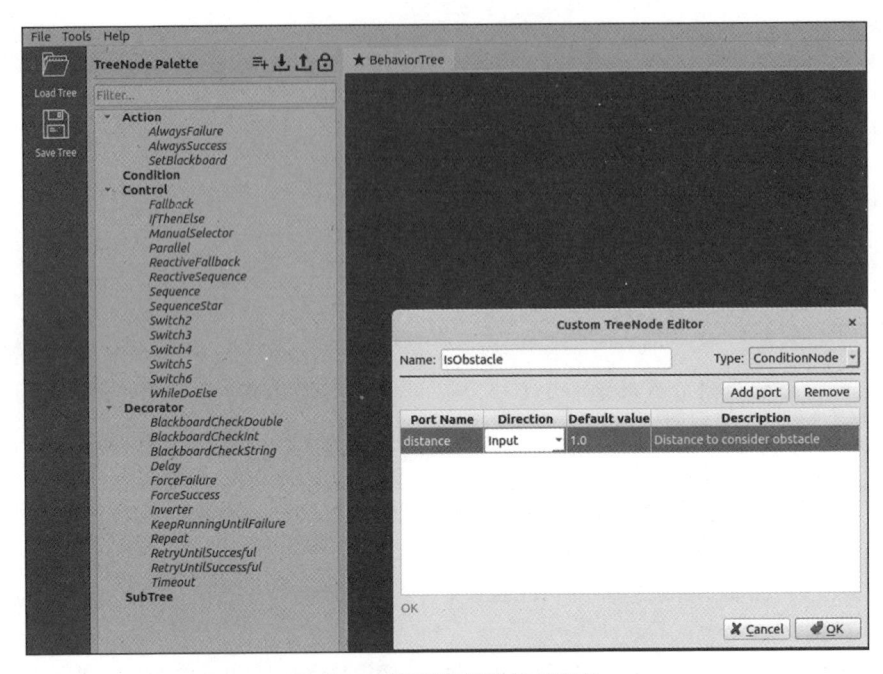

图 6.11 是否障碍物节点定义

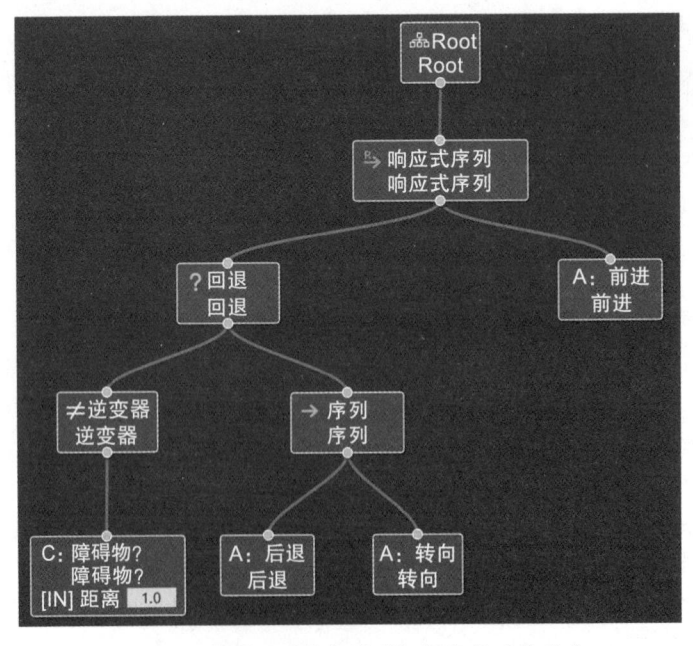

图 6.12 实现"碰撞与前进"任务的动作节点

```
behavior_tree_xml/bumpgo.xml
<?xml version="1.0"?>
<root main_tree_to_execute="BehaviorTree">
    <BehaviorTree ID="BehaviorTree">
        <ReactiveSequence>
            <Fallback>
                <Inverter>
                    <Condition ID="IsObstacle" distance="1.0"/>
                </Inverter>
                <Sequence>
                    <Action ID="Back"/>
                    <Action ID="Turn"/>
                </Sequence>
            </Fallback>
            <Action ID="Forward"/>
        </ReactiveSequence>
    </BehaviorTree>
    <TreeNodesModel>
        <Action ID="Back"/>
        <Action ID="Forward"/>
        <Condition ID="IsObstacle">
            <input_port default="1.0" name="distance">Dist to consider obst</input_port>
        </Condition>
        <Action ID="Turn"/>
    </TreeNodesModel>
</root>
```

行为树的 XML 定义很简单，分为两个部分：

- 行为树（BehaviorTree）：它定义了树结构。XML 标记与指定的行为树节点类型相匹配，子节点位于其父节点内。
- 节点模型（TreeNodesModel）：它定义了我们创建的自定义节点，指定其输入和输出端口。

这种结构的有效替代方案是忽略节点模型（TreeNodesModel），直接使用自定义行为树节点的名称：

```
<?xml version="1.0"?>
<root main_tree_to_execute="BehaviorTree">
    <BehaviorTree ID="BehaviorTree">
        <ReactiveSequence>
            <Fallback>
                <Inverter>
                    <IsObstacle distance="1.0"/>
                </Inverter>
                <Sequence>
                    <Back/>
                    <Turn/>
                </Sequence>
            </Fallback>
            <Forward/>
        </ReactiveSequence>
    </BehaviorTree>
</root>
```

6.2.2 行为树节点实现

我们将使用 Behavior Trees 库中的 behaviortree.CPP[⊖]，这在 ROS/ROS2 中是相当标准的。让我们看看 Forward 节点实现，以了解实现 BT 节点是多么简单：

⊖ https://www.behaviortree.dev。

```
include/mr2_bt_bumpgo/Forward.hpp
```
```cpp
class Forward : public BT::ActionNodeBase
{
public:
  explicit Forward(
    const std::string & xml_tag_name,
    const BT::NodeConfiguration & conf);

  BT::NodeStatus tick();

  static BT::PortsList providedPorts()
  {
    return BT::PortsList({});
  }

private:
  rclcpp::Node::SharedPtr node_;
  rclcpp::Time start_time_;
  rclcpp::Publisher<geometry_msgs::msg::Twist>::SharedPtr vel_pub_;
};
```

如前面的代码所示，当创建一个行为树时，对于行为树中出现的每个节点类，都会构建一个实例。动作节点会继承自 BT::ActionNodeBase，需要实现三个抽象方法并设置构造函数的参数：

- 构造函数接收 XML 中 name 字段的内容（这是可选的），以及包含指向所有树节点共享的黑板指针的 BT::NodeConfiguration，其中还包括其他内容。
- halt 方法在树完成执行时被调用，用于执行节点需要的任何清理工作。我们将其定义为 void，因为它是一个纯虚拟方法。
- tick 方法实现了我们在本章中已经描述的执行操作。
- 一个返回节点的端口的静态方法。在这种情况下，前进（Forward）节点不涉及端口，因此返回一个空的端口列表。

类的定义也很简单：

```
src/mr2_bt_bumpgo/Forward.cpp
```
```cpp
Forward::Forward(
  const std::string & xml_tag_name,
  const BT::NodeConfiguration & conf)
: BT::ActionNodeBase(xml_tag_name, conf)
{
  config().blackboard->get("node", node_);

  vel_pub_ = node_->create_publisher<geometry_msgs::msg::Twist>("/output_vel", 100);
}

BT::NodeStatus
Forward::tick()
{
  geometry_msgs::msg::Twist vel_msgs;

  vel_msgs.linear.x = 0.3;
  vel_pub_->publish(vel_msgs);

  return BT::NodeStatus::RUNNING;
}

}  // namespace br2_bt_bumpgo
```

```
src/mr2_bt_bumpgo/Forward.cpp
```
```cpp
#include "behaviortree_cpp_v3/bt_factory.h"
BT_REGISTER_NODES(factory)
{
  factory.registerNodeType<br2_bt_bumpgo::Forward>("Forward");
}
```

- 在构造函数中，在调用基类的构造函数之后，我们将获得黑板的 ROS2 节点指针。在创建树时，ROS2 节点的指针被写入到黑板，并使用键 node 使其对任何需要访问它的行为树节点可用，以便创建发布者和订阅者、获取时间或与 ROS2 相关的任何任务。
- tick 方法目的很明显：每次节点被执行时，它都会发布一个前进的速度消息，并返回运行中（RUNNING）状态。
- 在前面代码的最后部分，我们将此类注册为实现前进（Forward）节点的类。这部分将在创建树时使用。

在分析了前进（Forward）节点之后，其余的节点的实现与之类似。让我们看一些特殊之处：

- 转向（Turn）节点执行了 3s 的任务，因此它保存了第一次执行的时间戳，该时间戳是可识别的，因为其状态仍为空闲（IDLE）：

```
src/mr2_bt_bumpgo/Turn.cpp
```
```cpp
BT::NodeStatus
Turn::tick()
{
  if (status() == BT::NodeStatus::IDLE) {
    start_time_ = node_->now();
  }

  geometry_msgs::msg::Twist vel_msgs;
  vel_msgs.angular.z = 0.5;
  vel_pub_->publish(vel_msgs);

  auto elapsed = node_->now() - start_time_;

  if (elapsed < 3s) {
    return BT::NodeStatus::RUNNING;
  } else {
    return BT::NodeStatus::SUCCESS;
  }
}
```

- 是否障碍物（IsObstacle）节点保存激光器读数并将其与输入端口上设置的距离进行比较：

```
src/mr2_bt_bumpgo/IsObstacle.cpp
```
```cpp
void
IsObstacle::laser_callback(sensor_msgs::msg::LaserScan::UniquePtr msg)
{
  last_scan_ = std::move(msg);
}
```

```
src/mr2_bt_bumpgo/isObstacle.cpp
```
```cpp
BT::NodeStatus
IsObstacle::tick()
{
  double distance = 1.0;
  getInput("distance", distance);

  if (last_scan_->ranges[last_scan_->ranges.size() / 2] < distance) {
    return BT::NodeStatus::SUCCESS;
  } else {
    return BT::NodeStatus::FAILURE;
  }
}
```

每个行为树节点都将被编译为一个单独的库。稍后我们将看到，在创建包含它们的行为树时，我们可以将这些库作为插件加载，以便快速定位自定义行为树节点的实现。

```
CMakeLists.txt
```
```cmake
add_library(br2_forward_bt_node SHARED src/br2_bt_bumpgo/Forward.cpp)
add_library(br2_back_bt_node SHARED src/br2_bt_bumpgo/Back.cpp)
add_library(br2_turn_bt_node SHARED src/br2_bt_bumpgo/Turn.cpp)
add_library(br2_is_obstacle_bt_node SHARED src/br2_bt_bumpgo/IsObstacle.cpp)

list(APPEND plugin_libs
  br2_forward_bt_node
  br2_back_bt_node
  br2_turn_bt_node
  br2_is_obstacle_bt_node
)

foreach(bt_plugin ${plugin_libs})
  ament_target_dependencies(${bt_plugin} ${dependencies})
  target_compile_definitions(${bt_plugin} PRIVATE BT_PLUGIN_EXPORT)
endforeach()

install(TARGETS
  ${plugin_libs}
  ARCHIVE DESTINATION lib
  LIBRARY DESTINATION lib
  RUNTIME DESTINATION lib/${PROJECT_NAME}
)
```

6.2.3 运行行为树

运行行为树是简单的。一个程序应该构建一棵树，并开始对其根进行执行，直到它返回成功（SUCCESS）状态。行为树是通过使用 BehaviorTreeFactory 创建的，它需要指定一个 XML 文件或指定包含 XML 文件路径的字符串。BehaviorTreeFactory 需要加载自定义节点的库作为插件，并需要在行为树节点之间共享黑板（blackboard）。

为了将行为树与 ROS2 集成，需要创建一个 ROS2 节点并将其存放在黑板上。如前所示，行为树节点可以从黑板中提取它，以便创建某个服务或动作的发布者/订阅者或客户端/服务端。与树根的 tick 一起，spin_some 负责管理消息到达 ROS2 节点。

看一下具体负责创建和执行树的程序的实现：

```
src/mr2_bt_bumpgo/isObstacle.cpp

int main(int argc, char * argv[])
{
  rclcpp::init(argc, argv);

  auto node = rclcpp::Node::make_shared("patrolling_node");

  BT::BehaviorTreeFactory factory;
  BT::SharedLibrary loader;

  factory.registerFromPlugin(loader.getOSName("br2_forward_bt_node"));
  factory.registerFromPlugin(loader.getOSName("br2_back_bt_node"));
  factory.registerFromPlugin(loader.getOSName("br2_turn_bt_node"));
  factory.registerFromPlugin(loader.getOSName("br2_is_obstacle_bt_node"));

  std::string pkgpath = ament_index_cpp::get_package_share_directory("br2_bt_bumpgo");
  std::string xml_file = pkgpath + "/behavior_tree_xml/bumpgo.xml";

  auto blackboard = BT::Blackboard::create();
  blackboard->set("node", node);
  BT::Tree tree = factory.createTreeFromFile(xml_file, blackboard);

  auto publisher_zmq = std::make_shared<BT::PublisherZMQ>(tree, 10, 1666, 1667);

  rclcpp::Rate rate(10);

  bool finish = false;
  while (!finish && rclcpp::ok()) {
    finish = tree.rootNode()->executeTick() != BT::NodeStatus::RUNNING;

    rclcpp::spin_some(node);
    rate.sleep();
  }

  rclcpp::shutdown();
  return 0;
}
```

- 在主函数的开始，我们创建了一个通用的 ROS2 节点，然后将其插入到黑板中。这就是我们在 Forward 中从黑板中提取出来创建速度消息发布者的节点。
- 行为树是由 BT::BehaviorTreeFactory 工厂类从一个 XML 文件、将要实现的行为树动作节点以及一个黑板对象创建的。
  1）正如我们将在下面看到的，每个行为树节点将被编译成独立的库。加载器（loader）对象从系统中找到库并将行为树节点作为插件加载。我们之前在行为树节点定义中见到的 BT_REGISTER_NODES 宏允许将行为树节点名称与其在库中的实现连接起来。

```
BT::BehaviorTreeFactory factory;
BT::SharedLibrary loader;

factory.registerFromPlugin(loader.getOSName("br2_forward_bt_node"));
```

  2）ament_index_cpp 包中的 get_package_share_directory 函数可以获取已安装软件包的完整路径，以便读取其中的任何文件。请记住，这是一个包含在包的依赖项中的软件包。

```
std::string pkgpath = ament_index_cpp::get_package_share_directory(
  "br2_bt_bumpgo");
std::string xml_file = pkgpath + "/behavior_tree_xml/forward.xml";
```

3）最后，在创建了黑板并将其中的共享指针插入 ROS2 节点后，工厂构建了执行树。

```
auto blackboard = BT::Blackboard::create();
blackboard->set("node", node);
BT::Tree tree = factory.createTreeFromFile(xml_file, blackboard);
```

4）为了在运行时调试行为树，创建一个 PublisherZMQ 对象，该对象用来发布所有必要的信息。在创建它时，需要指定树、每秒最大消息数以及要使用的网络端口。

```
auto publisher_zmq = std::make_shared<BT::PublisherZMQ>(
    tree, 10, 1666, 1667);
```

- 在最后部分，树的根以 10Hz 被执行，同时树在处理节点中的任何挂起工作（例如传递到订阅者的消息）时返回运行中（RUNNING）状态。

编译完成后，执行模拟器和节点并运行程序。机器人应该会向前移动。

```
$ ros2 launch br2_tiago sim.launch.py
```

```
$ ros2 run br2_bt_bumpgo bt_bumpgo --ros-args -r input_scan:=/scan_raw -r output_vel:=/key_vel -p use_sim_time:=true
```

在程序执行期间，可以使用 Groot 来监视行为树的状态，了解哪些节点正在被执行以及它们返回的值。只需要启动 Groot，选择监视器（Monitor）而不是编辑器（Editor）。一旦按下连接后，就可监视执行，如图 6.13 所示。

6.2.4 测试行为树节点

在这个包的 tests 目录中包含了两种类型的测试，这些测试在项目的开发过程中都非常有用。

第一种测试是单独测试每个节点，运行只包含一种节点类型的行为树，以查看它们在隔离环境中是否正常工作。我们只列举了对行为树前进（Forward）节点的验证：

```
tests/bt_forward_main.cpp
factory.registerFromPlugin(loader.getOSName("br2_forward_bt_node"));

std::string xml_bt =
  R"(
<root main_tree_to_execute = "MainTree" >
    <BehaviorTree ID="MainTree">
        <Forward />
    </BehaviorTree>
</root>)";

auto blackboard = BT::Blackboard::create();
blackboard->set("node", node);
BT::Tree tree = factory.createTreeFromText(xml_bt, blackboard);

rclcpp::Rate rate(10);
```

```
tests/bt_forward_main.cpp
bool finish = false;
while (!finish && rclcpp::ok()) {
  finish = tree.rootNode()->executeTick() != BT::NodeStatus::RUNNING;

  rclcpp::spin_some(node);
  rate.sleep();
}
```

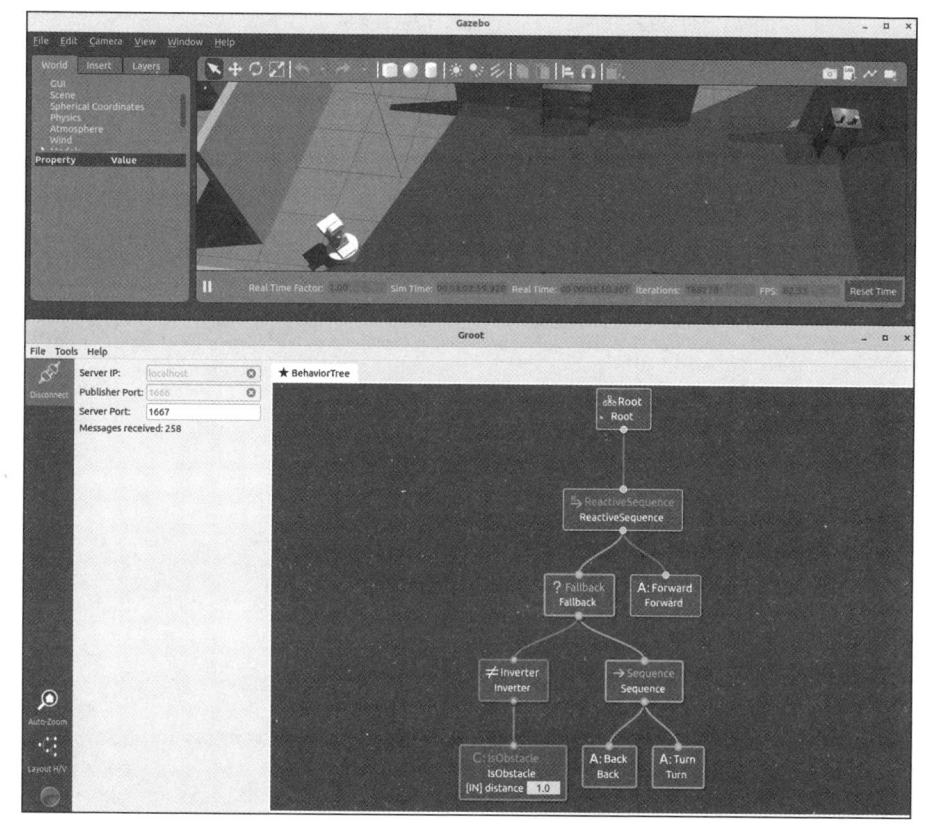

图 6.13 使用 Groot 监视行为树的执行（见彩插）

启动模拟器并运行：

```
$ build/br2_bt_bumpgo/tests/bt_forward --ros-args -r input_scan:=/scan_raw -r
output_vel:=/key_vel -p use_sim_time:=true
```

检查机器人是否会一直向前移动。对其他行为树节点执行同样的测试。

第二种测试是上一章中推荐的使用 GoogleTest 进行的测试。很容易定义一个 ROS2 节点，使其记录已发送到速度消息主题的速度。

```cpp
// tests/bt_action_test.cpp
class VelocitySinkNode : public rclcpp::Node
{
public:
  VelocitySinkNode()
  : Node("VelocitySink")
  {
    vel_sub_ = create_subscription<geometry_msgs::msg::Twist>(
      "/output_vel", 100, std::bind(&VelocitySinkNode::vel_callback, this, _1));
  }

  void vel_callback(geometry_msgs::msg::Twist::SharedPtr msg)
  {
    vel_msgs_.push_back(*msg);
  }

  std::list<geometry_msgs::msg::Twist> vel_msgs_;

private:
  rclcpp::Subscription<geometry_msgs::msg::Twist>::SharedPtr vel_sub_;
};
```

可以多次执行一棵行为树,以检查发送的速度是否正确:

```cpp
// tests/bt_action_test.cpp
TEST(bt_action, forward_btn)
{
  auto node = rclcpp::Node::make_shared("forward_btn_node");
  auto node_sink = std::make_shared<VelocitySinkNode>();

  // Creation the Behavior Tree only with the Forward BT node

  rclcpp::Rate rate(10);
  auto current_status = BT::NodeStatus::FAILURE;
  int counter = 0;
  while (counter++ < 30 && rclcpp::ok()) {
    current_status = tree.rootNode()->executeTick();
    rclcpp::spin_some(node_sink);
    rate.sleep();
  }

  ASSERT_EQ(current_status, BT::NodeStatus::RUNNING);
  ASSERT_FALSE(node_sink->vel_msgs_.empty());
  ASSERT_NEAR(node_sink->vel_msgs_.size(), 30, 1);

  geometry_msgs::msg::Twist & one_twist = node_sink->vel_msgs_.front();

  ASSERT_GT(one_twist.linear.x, 0.1);
  ASSERT_NEAR(one_twist.angular.z, 0.0, 0.0000001);
}
```

在这种情况下,在一个行为树树根被执行了 30 次后,查看节点是否仍然返回运行中(RUNNING)状态、是否已发布了 30 条与速度相关的消息,并且速度是正确的(它们使机器人向前移动)。我们可以检查所有的速度,但我们只对发布的第一条速度进行了测试。

检查其他节点的测试。对于转向(Turn)节点和后退(Back)节点,我们检查了它们在返回成功状态之前是否在适当的时间执行。对于是否障碍物(IsObstacle)节点,我们创建了合成的激光读数,以查看输出在所有情况下是否正确。

6.3 使用行为树进行巡逻

在这一节中，我们将处理一个更复杂和具有挑战性的项目。我们之前说过，行为树动作节点有助于控制其他子系统。在上一节的项目中，我们以一种非常基础的方式处理了感知信息并发送速度信息。在本节中，我们将执行一个项目，其中行为树将控制更复杂的子系统，如 Nav2 导航子系统和主动视觉子系统。

本节项目的目标是使机器人在 Gazebo㊀ 环境的模拟房屋中巡逻：

- 机器人在房屋中巡逻三个路标（见图 6.14）。到达每个路标时，机器人会围绕自身转动一段时间以感知周围环境。
- 在机器人从一个路标到另一个路标的过程中，机器人会感知并跟踪检测到的对象。
- 机器人会跟踪（模拟）其电池电量。当电量低时，它会去一个充电点进行几秒钟的充电。

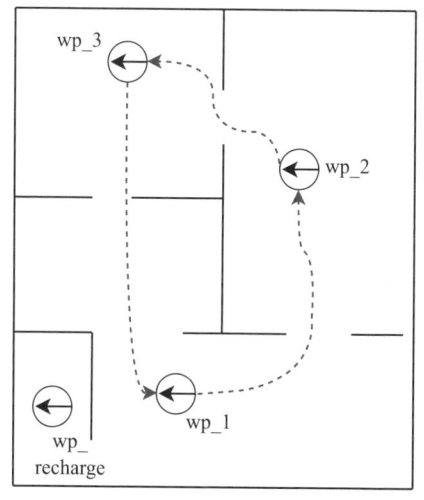

图 6.14 模拟房屋中的路标和巡逻过程中经过的路径

由于我们正在使用像 Nav2 这样复杂且重要的子系统，它是 ROS2 中的导航系统，因此我们首先在 6.3.1 节中描述它。6.3.2 节描述了为特定机器人和环境设置 Nav2 的步骤，可以跳过这个部分，因为 br2_navigation 包已经包含了针对模拟 Tiago 场景在房子中的环境地图和配置文件。接下来的部分将专注于如何实现行为树和巡逻节点。

6.3.1 Nav2 介绍

Nav2[3] 是 ROS2 导航系统，它是模块化、可配置和可扩展的。与 ROS 的前身一样，它旨在成为最广泛使用的导航软件，因此支持主要的机器人类型：全向驱动、差分驱动㊁、腿式类型和 Ackermann（类似汽车）类型㊂等，同时允许合并来自激光器和 3D 相机等信息。Nav2 集成了多个用于本地和全局导航的插件，并允许轻松使用自定义插件。

Nav2 的输入包括：符合 REP-105 标准的 TF 变换、地图㉔以及任何相关的传感

㊀ Gazebo 是一个用于仿真机器人和其他物理系统的开源三维模拟器。——译者注
㊁ 两轮。——译者注
㊂ 通常有四个轮子。——译者注
㉔ 如果使用静态成本地图。

器数据源。它还需要导航逻辑（以行为树 XML 文件编码的形式呈现），并在需要时对其进行特定问题的调整。Nav2 的输出是发送到机器人底座的速度。

在模拟房屋中的路标和巡逻期间所遵循的路径如图 6.15 所示。让我们描述一下图中出现的每个组件。

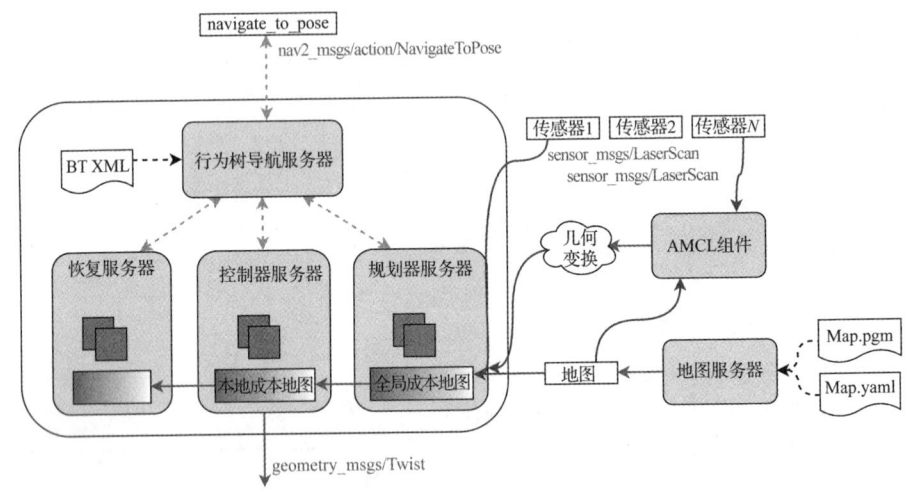

图 6.15 在模拟房屋中的路标和巡逻期间所遵循的路径

- **地图服务器**：该组件从两个文件中读取地图，并将其作为 nav_msgs/msg/OccupancyGrid 发布，节点在内部将其处理为 costmap2D。Nav2 中的地图是网格，其单元格编码空间是空闲（0）、未知（255）或占用（254）。在 1 到 253 之间的数值代表了穿越该区域不同的占用程度或成本。图 6.16 b 展示了控制器服务器使用的原始地图和本地成本地图（以 costmap2D 编码）。

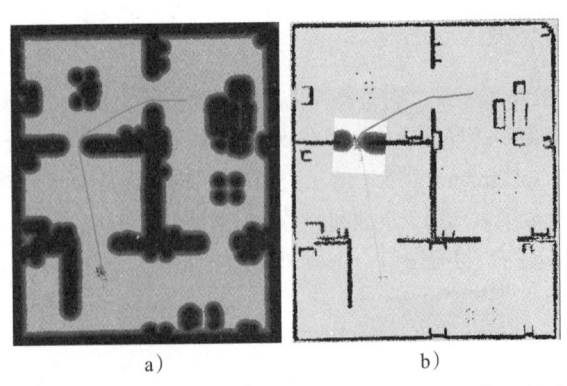

图 6.16 Nav2 组件使用的 2D 成本地图：a）规划器服务器使用的全局成本地图；b）控制器服务器使用的原始地图和本地成本地图

- **AMCL**：该组件实现了一种基于自适应蒙特卡洛（AMCL）[9]的定位算法。它使用传感器信息，并使用激光和地图的读数来计算机器人的位置。输出是一个几何变换，指示了机器人的位置。由于每个坐标系都不应有两个父坐标系，因此该组件计算并发布 map→odom 变换，而不是 map→base_footprint 变换。
- **规划器服务器**：该组件的功能是计算从起点到终点的路线。它将目的地、机器人的当前位置和环境地图作为输入。规划器服务器根据原始地图构建一个成本地图，其中墙壁被机器人的半径和安全余量加粗。这样，机器人可以使用空闲空间（或成本较低的空间）来计算路线，规划器服务器使用的全局成本地图如图 6.16a 所示。路线规划和成本地图更新算法作为插件加载。像接下来的两个组件一样，规划器服务器通过 ROS2 动作接收请求。
- **控制器服务器**：该组件接收规划器服务器计算出的路线，并向机器人的基座发布发送的速度。它使用机器人周围的本地成本地图（见图 6.16b），其中附近的障碍物被编码，并用算法（作为插件加载）来计算速度。
- **恢复服务器**：该组件具有几种有用的恢复策略，如果机器人迷路、被困或无法计算到达终点的路线，这些策略包括转弯、清除成本地图、缓慢移动等。
- **行为树导航服务器**：该组件是负责协调其他导航组件的组件。它以 ROS2 动作的形式接收导航请求。动作名称是 navigate_to_pose，类型是 nav2_msgs/action/NavigateToPose。因此，如果我们想让机器人从一个点移动到另一个点，我们就必须使用这个 ROS2 动作。查看这个动作：

```
ros2 interface show nav2_msgs/action/NavigateToPose
#goal definition
geometry_msgs/PoseStamped pose
string behavior_tree
---
#result definition
std_msgs/Empty result
---
geometry_msgs/PoseStamped current_pose
builtin_interfaces/Duration navigation_time
int16 number_of_recoveries
float32 distance_remaining
```

1）请求部分包括目标位置，以及可选的用于在此动作中替代默认的自定义行为树。这个最后的功能允许进行特殊请求，这些请求不是正常的导航行为，例如跟随移动的物体或以特定方式接近障碍物。
2）动作完成时的结果。
3）在导航过程中，机器人持续返回当前位置、到目标的距离以及统计数据，例如导航时间或从不良情况中恢复的次数。

行为树导航器使用行为树来编排机器人的导航。行为树节点向 Nav2 的其他组件发出请求，以便它们执行各自的任务。

当该组件接收一个导航动作时，它会开始执行一个行为树，示例如图 6.17 所示。Nav2 的默认行为树要复杂得多，包括恢复的调用，但图中的行为树很好地说明了行为树导航服务器如何使用它们。首先，通过 ROS2 动作到达的目标被写入黑板上。ComputePathToPose 使用这个目标来调用规划器服务器的动作，该动作返回到达目标的路线。这个路线是这个行为树节点的输出，也是 FollowPath 行为树节点的输入，并将后者发送给控制器服务器。

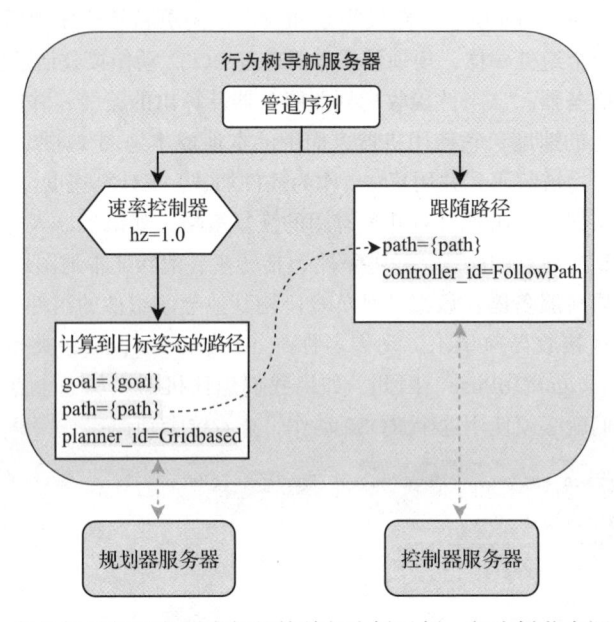

图 6.17 行为树导航服务器内部的简单行为树示例，行为树节点调用 ROS2 动作以协调其他 Nav2 组件

要使用 Nav2，只需要安装包含 Nav2 的软件包即可：

```
$ sudo apt install ros-humble-navigation2 ros-humble-nav2-bringup ros-humble-turtlebot3*
```

在 br2_navigation 软件包中，我们已经准备好了模拟 Tiago 机器人在家庭场景中导航所需的启动器、地图和配置文件。让我们进行导航测试：

- 启动模拟器：

```
$ ros2 launch br2_tiago sim.launch.py
```

- 启动导航器：

```
$ ros2 launch br2_navigation tiago_navigation.launch.py
```

- 打开 RViz2 并显示，Nav2 在执行中如图 6.18 所示：
  1）TF：显示机器人。观察 map→odom 的变换。
  2）地图：显示 /map 主题，其 QoS 是可靠且瞬态本地的。
  3）全局成本地图：显示 /global_costmap/costmap 主题，默认 QoS 为可靠和易失性。
  4）局部成本地图：显示 /local_costmap/costmap 主题，默认 QoS。
  5）激光扫描：显示它与障碍物的匹配情况。
  6）显示 AMCL 粒子很有趣。每个粒子都是关于机器人位置的假设。最终的机器人位置是所有这些粒子的平均值。箭头群体越集中，机器人定位越好。它在 /particlecloud 消息主题中，其 QoS 为最努力与易失性。

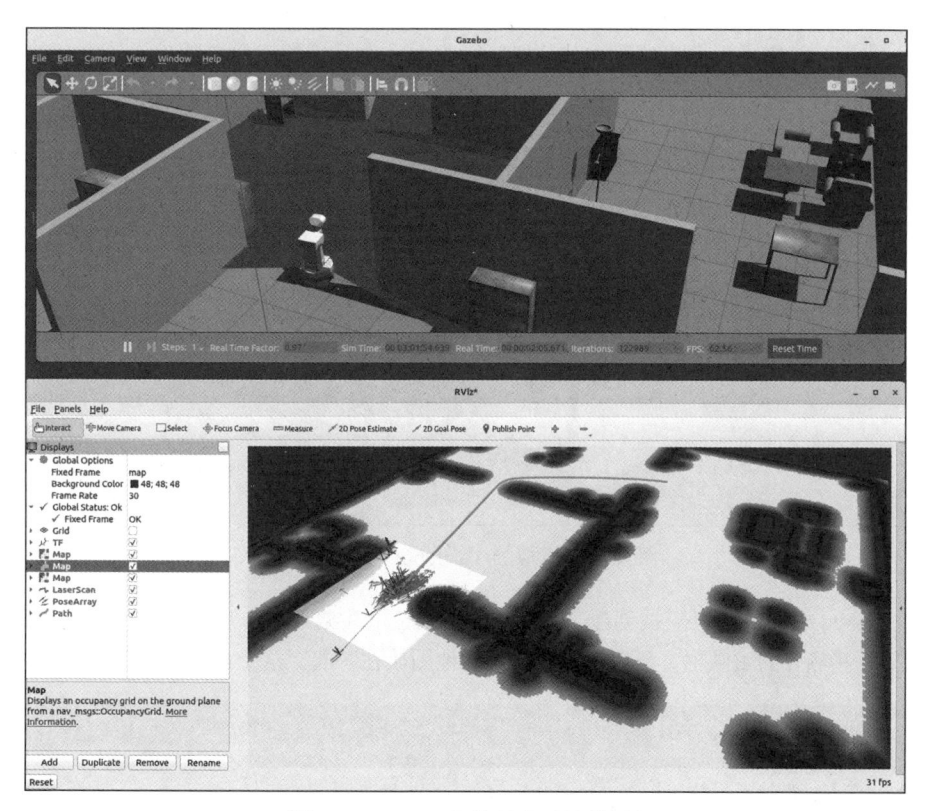

图 6.18　Nav2 在执行中（见彩插）

- 使用 "2D Goal Pose" 按钮向机器人发送目标位置命令。

- 在获取地图位置时,使用"Publish Point"按钮。然后在地图上的任何位置单击。此位置将被发布到 /clicked_point 消息主题。

6.3.2 设置 Nav2 参数

本节介绍了为新环境和特定机器人设置 Nav2 的过程。如果你使用的是 br2_navigation 软件包,则可以跳过此步骤,因为其已包含了使模拟的 Tiago 机器人在家庭场景中导航所需的一切。如果你想使用其他场景或其他机器人,请继续阅读。

如果通过软件包安装了 Nav2,则它位于 /opt/ros/humble/ 目录下。特别是 /opt/ros/humble/share/nav2_bringup 目录包含了 Nav2 的启动器、地图和参数,该参数用于在默认情况下提供模拟 Turtlebot3。你可以通过输入以下命令来启动它:

```
$ ros2 launch nav2_bringup tb3_simulation_launch.py
```

它启动了一个在小型环境中的 Turtlebot3 模拟(见图 6.19)。使用 "2D Pose Estimate" 按钮将机器人放置在其位置,因为在这之前导航不会被激活。

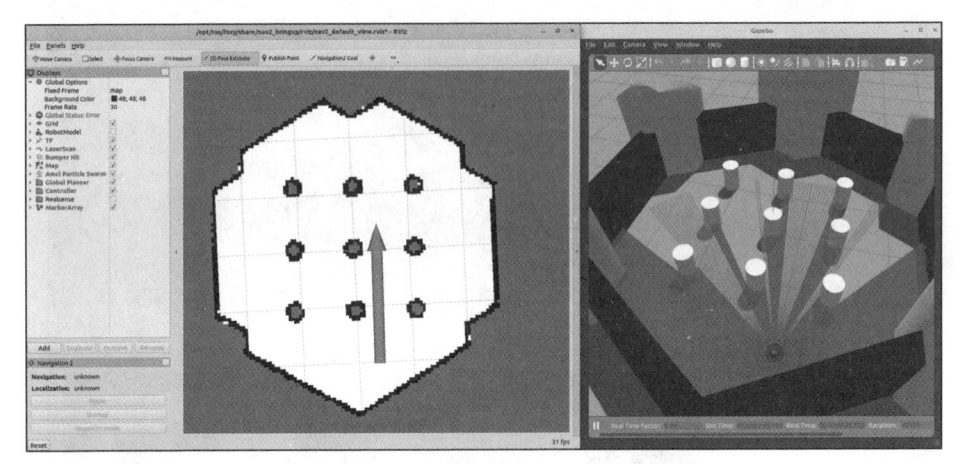

图 6.19 模拟的 Turtlebot3(见彩插)

Tiago 模拟的包已经被创建,它从 nav2_bringup 中复制了一些元素,因为启动器需要一些额外的重映射,所以将配置文件和地图放在一起。这个包具有以下结构:

```
Package br2_navigation

br2_navigation/
├── CMakeLists.txt
├── launch
│   ├── navigation_launch.py
│   └── tiago_navigation.launch.py
├── maps
│   ├── home.pgm
│   └── home.yaml
```

```
Package br2_navigation
├── package.xml
└── params
    ├── mapper_params_online_async.yaml
    └── tiago_nav_params.yaml
```

首先看看如何映射环境。我们将使用 slam_toolbox 软件包。我们将使用一个自定义的参数文件来指定特定的消息主题和坐标系：

```
params/mapper_params_online_async.yaml
# ROS Parameters
odom_frame: odom
map_frame: map
base_frame: base_footprint
scan_topic: /scan_raw
mode: mapping #localization
```

运行这些命令，每个命令都在不同的终端中执行：

- 在家庭场景的模拟环境中启动 Tiago 机器人。

```
$ ros2 launch br2_tiago sim.launch.py
```

- 使用 RViz2 可视化映射进度，定位和地图渲染（SLAM）见图 6.20。

```
$ rviz2 --ros-args -p use_sim_time:=true
```

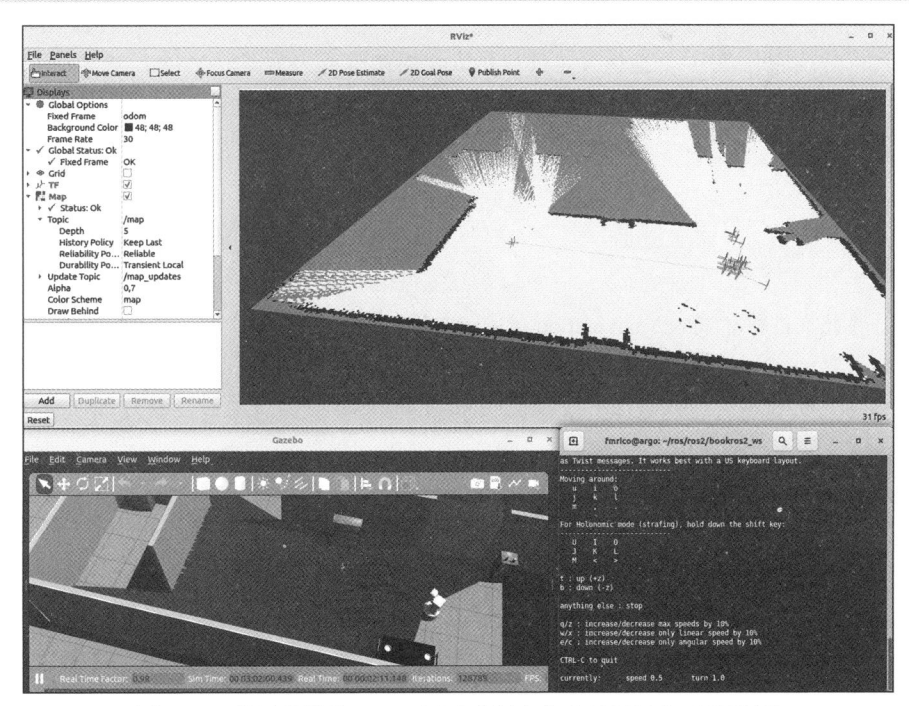

图 6.20 使用模拟的 Tiago 同时进行定位和地图渲染（见彩插）

- 启动 SLAM 节点。它会将正在构建的地图发布到 /map。

```
$ ros2 launch slam_toolbox online_async_launch.py params_file:=[Full path
to bookros2_ws/src/book_ros2/br2_navigation/params/mapper_params_online_async
.yaml] use_sim_time:=true
```

- 启动地图保存服务器。此节点将订阅 /map，当收到请求时将其保存到磁盘。

```
$ ros2 launch nav2_map_server map_saver_server.launch.py
```

- 运行远程操作器使机器人沿着场景移动。
- 执行代码，在不同的终端中运行这些命令。

```
$ ros2 run teleop_twist_keyboard teleop_twist_keyboard --ros-args --remap
/cmd_vel:=/key_vel -p use_sim_time:=true
```

一旦开始使用远程操作器使机器人在舞台上移动，就启动 RViz2 并检查地图的构建情况。当地图完成时，请求地图保存服务器将地图保存到磁盘：

```
$ ros2 run nav2_map_server map_saver_cli --ros-args -p use_sim_time:=true
```

请注意，当使用真实机器人进行地图渲染或导航时，启动器和节点中的 use_sim_time 参数必须为 false。

此时，将创建两个文件。一个是 PGM 图像文件（如果你需要进行任何修复，就可以修改它），另一个是包含足够信息以将图像解释为地图的 YAML 文件。请记住，如果修改了文件名，则这个 YAML 文件也应修改：

```
image: home.pgm
mode: trinary
resolution: 0.05
origin: [-2.46, -13.9, 0]
negate: 0
occupied_thresh: 0.65
free_thresh: 0.25
```

将此文件移动到 br2_navigation 包中，并继续进行下一步设置。在这一步中，需要修改从 nav2_bringup 中复制的启动器。使用包含启动器的 tiago_navigation.launch 来启动导航和定位。我们没有直接使用 nav2_bringup 中的启动器，因为 navigation.launch 需要一些额外的重映射。

```
br2_navigation/launch/navigation_launch.py

remappings = [('/tf', 'tf'),
('/tf_static', 'tf_static'),
('/cmd_vel', '/nav_vel')
]
```

关于参数文件，首先从 nav2_bringup 包中的参数文件开始。以下是一些配置的详细信息：

- 首先，最重要的是将所有包含传感器消息主题的参数设置为正确的值，并确保所有的坐标系都在我们的机器人中存在且正确。
- 如果初始位置已知，则在 AMCL 配置中设置它。如果你以与开始映射时相同的姿态启动机器人，则是（0, 0, 0）位置。

```
br2_navigation/params/tiago_nav_params

amcl:
  ros__parameters:
    scan_topic: scan_raw
    set_initial_pose: true
    initial_pose:
      x: 0.0
      y: 0.0
      z: 0.0
      yaw: 0.0
```

- 根据机器人的能力设置速度和加速度：

```
br2_navigation/params/tiago_nav_params

controller_server:
  ros__parameters:
    use_sim_time: False
    FollowPath:
      plugin: "dwb_core::DWBLocalPlanner"
      min_vel_x: 0.0
      min_vel_y: 0.0
      max_vel_x: 0.3
      max_vel_y: 0.0
      max_vel_theta: 0.5
      min_speed_xy: 0.0
      max_speed_xy: 0.5
      min_speed_theta: 0.0
      acc_lim_x: 1.5
      acc_lim_y: 0.0
      acc_lim_theta: 2.2
      decel_lim_x: -2.5
      decel_lim_y: 0.0
      decel_lim_theta: -3.2
```

- 将机器人的半径设置为膨胀墙壁和障碍物的值○，并设置一个用于决定导航距离的缩放因子。这些设置由膨胀层（inflation_layer）成本地图插件控制，适用于本地和全局成本地图：

```
br2_navigation/params/tiago_nav_params

local_costmap:
  local_costmap:
    ros__parameters:
      robot_radius: 0.3
      plugins: ["voxel_layer", "inflation_layer"]
      inflation_layer:
        plugin: "nav2_costmap_2d::InflationLayer"
        cost_scaling_factor: 3.0
        inflation_radius: 0.55
```

○ 通过增加机器人半径的大小，可以确保在路径规划时，机器人会在考虑到膨胀后的障碍物，并在周围留有足够的空间，避免碰撞或太靠近障碍物。这有助于提高导航的安全性。——译者注

6.3.3 计算图和行为树

巡逻项目的计算图（见图 6.21）由巡逻节点（patrolling_node）和属于两个被控制子系统的节点组成：Nav2 和在上一章中开发的主动视觉系统。

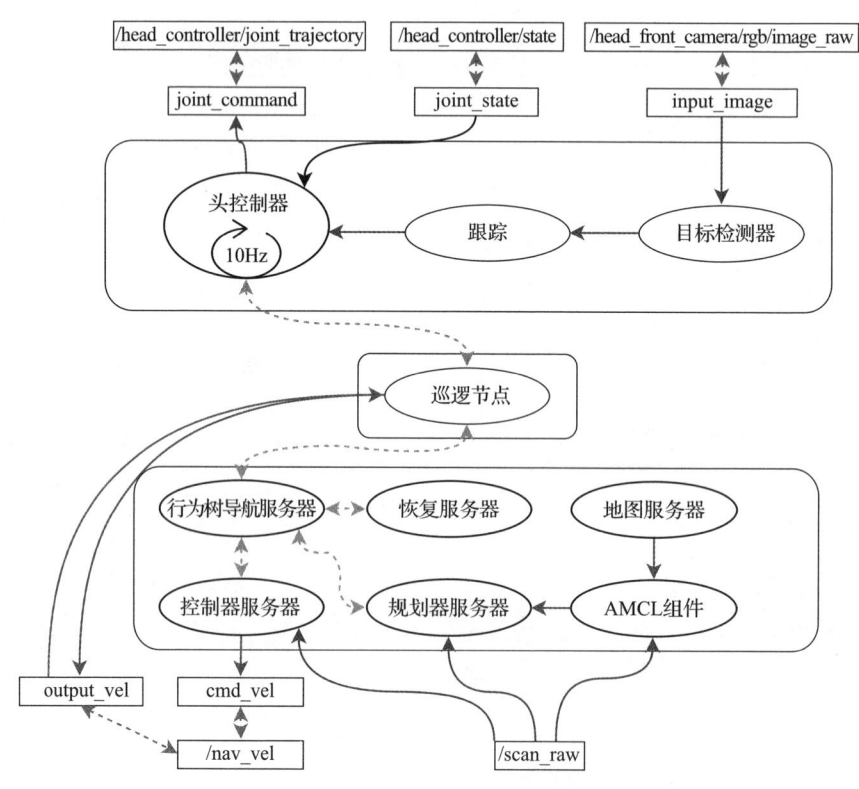

图 6.21 巡逻项目的计算图。为了描述清晰子系统已简化处理

- Nav2 使用 ROS2 动作进行控制，发送组成巡逻路线的目标姿态。
- 在导航期间，通过使用 ROS2 服务来激活主动视觉系统中的头控制器（HeadController）节点（一个生命周期节点，具有生命周期管理的功能）。
- 此外，在到达路标时，为了使机器人自行旋转，巡逻节点（patrolling_node）将直接向机器人底座发布速度信息。

计算图中的巡逻节点（patrolling_node）展现得非常简单。也许分析其包含的行为树更有意思，这是控制其控制逻辑的行为树。图 6.22 展示了巡逻项目的行为树的完整结构。让我们分析其中的每个动作和条件节点。

- **移动（Move）节点**：该节点负责通过 ROS2 动作向 Nav2 发送导航请求。导航目标通过其目标（goal）端口的输入端口接收，该目标端口包含一个 (x, y)

位置和一个 theta 方向的坐标。该节点返回运行中状态直到收到导航操作完成的通知，此时它将返回成功状态。返回失败状态的情况未被考虑，尽管这样做会很方便。

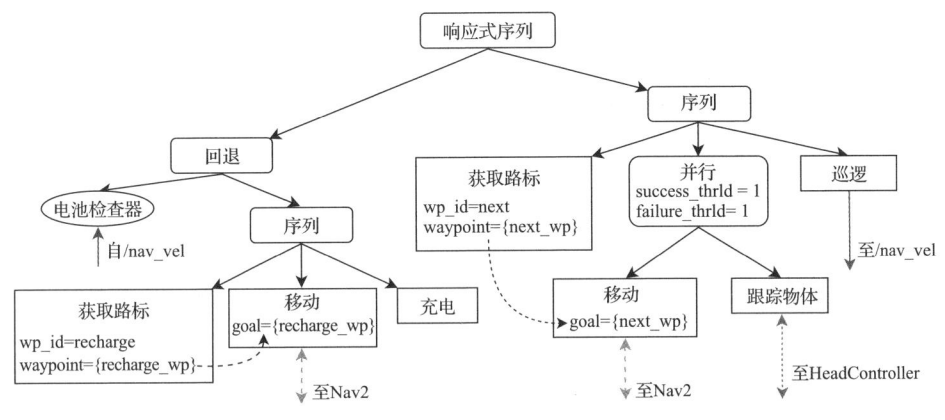

图 6.22 巡逻项目的行为树

- **获取路标（GetWaypoint）节点**：该节点用于获取移动节点使用的几何坐标。它有一个包含几何坐标的路标输出端口，并将其作为行为树节点的输入。获取路标（GetWaypoint）节点的输入是一个表示所需路标的 id 值。如果此输入是"充电"，则其输出是充电点的坐标。如果输入是"下一个"，则返回要导航到的下一个路标的几何坐标。该节点的存在是为了简化行为树，否则树的右侧分支需要重复三次，每个路标执行一次。第二个原因是将选择目标点的任务委派给另一个行为树节点，从而简化移动（Move）节点，无需在内部维护所有路标的坐标。有许多其他替代方案，但这是一个相当清晰和可扩展的选择。
- **电池检查器（BatteryChecker）节点**：该节点模拟机器人的电池电量。它在黑板上维护着电池电量信息，随着时间的推移和机器人的移动而减少电量（这就是为什么它订阅了命令执行速度的消息主题）。如果电池电量降到一定水平，它就会返回失败状态，否则返回成功状态。
- **巡逻（Patrol）节点**：该节点会简单地让机器人旋转一段时间以调整外部环境。当完成时，返回成功状态。
- **跟踪物体（TrackObjects）节点**：该节点始终返回运行中状态。当首次执行时，它会激活（如果尚未激活）头控制器（HeadController）节点。此节点与移动（Move）节点并行运行。配置并行（Parallel）控制节点，当两个节点中的一个（只能是移动节点）返回成功状态时，它认为其所有子节点的任务已完成，从而停止状态仍为运行中状态的节点。当跟踪物体（TrackObjects）节点收到停止信号时，它会禁用头控制器（HeadController）节点。

6.3.4 巡逻任务的实现

br2_bt_patrolling 包的结构与上一节中的结构类似：它在类定义和声明的地方分别实现了每个行为树节点。它有用于创建并执行该树的一个主程序，并且为每个实现行为树的节点提供了测试用例。

```
Package br2_bt_patrolling
br2_bt_patrolling/
├── behavior_tree_xml
│   └── patrolling.xml
├── cmake
│   └── FindZMQ.cmake
├── CMakeLists.txt
├── include
│   └── br2_bt_patrolling
│       ├── BatteryChecker.hpp
│       ├── ctrl_support
│       │   ├── BTActionNode.hpp
│       │   └── BTLifecycleCtrlNode.hpp
│       ├── GetWaypoint.hpp
│       ├── Move.hpp
│       ├── Patrol.hpp
│       ├── Recharge.hpp
│       └── TrackObjects.hpp
├── launch
│   └── patrolling.launch.py
├── package.xml
├── src
│   ├── br2_bt_patrolling
│   │   ├── BatteryChecker.cpp
│   │   ├── GetWaypoint.cpp
│   │   ├── Move.cpp
│   │   ├── Patrol.cpp
│   │   ├── Recharge.cpp
│   │   └── TrackObjects.cpp
│   └── patrolling_main.cpp
└── tests
    ├── bt_action_test.cpp
    └── CMakeLists.txt
```

从实现角度来看，最有趣的是两个类，它们简化了使用 ROS2 动作的行为树节点和激活生命周期节点的行为数节点。它们位于 include/br2_bt_patrolling/ctrl_support 目录中，并且以一种通用的方式实现，以便可以复用于其他项目。

BTActionNode 类是从 Nav2 中借用的，Nav2 中的行为树导航服务器使用它来控制其余的服务器。这是一个相当复杂的类，因为它考虑的情况比我们在这个项目中使用的要多得多，例如取消和重新发送动作。我们不打算深入讨论其实现细节。我建议查看官方 ROS2 页面上的 ROS2 动作教程，以了解更多信息。

如果行为树节点想要通过 ROS2 动作控制一个子系统，就需要继承 BTActionNode 类。让我们分析其接口及其派生类。原始注释将帮助我们理解它们的实用性：

```
include/br2_bt_patrolling/ctrl_support/BTActionNode.hpp
template<class ActionT, class NodeT = rclcpp::Node>
class BtActionNode : public BT::ActionNodeBase
{
```

```cpp
include/br2_bt_patrolling/ctrl_support/BTActionNode.hpp
public:
  BtActionNode(
    const std::string & xml_tag_name,
    const std::string & action_name,
    const BT::NodeConfiguration & conf)
  : BT::ActionNodeBase(xml_tag_name, conf), action_name_(action_name)
  {
    node_ = config().blackboard->get<typename NodeT::SharedPtr>("node");
    ...
  }

  // Could do dynamic checks, such as getting updates to values on the blackboard
  virtual void on_tick()
  {
  }

  // Called upon successful completion of the action. A derived class can override this
  // method to put a value on the blackboard, for example.
  virtual BT::NodeStatus on_success()
  {
    return BT::NodeStatus::SUCCESS;
  }

  // Called when a the action is aborted. By default, the node will return FAILURE.
  // The user may override it to return another value, instead.
  virtual BT::NodeStatus on_aborted()
  {
    return BT::NodeStatus::FAILURE;
  }

  // The main override required by a BT action
  BT::NodeStatus tick() override
  {
    ...
  }

  // The other (optional) override required by a BT action. In this case, we
  // make sure to cancel the ROS 2 action if it is still running.
  void halt() override
  {
    ...
  }
protected:
  typename ActionT::Goal goal_;
};
```

BTActionNode 类是一个模板类，因为每个动作都有不同的类型。例如，移动节点的动作类型是 nav2_msgs/action/NavigateToPose。该类还使用 ROS2 节点类型进行参数化，因为它也可以用生命周期节点实例化。

执行（tick）和中断（halt）方法由 BTActionNode 类处理，因此不应在派生类中被定义。可以在派生类中重写其他方法以执行某些操作，例如在动作完成或失败时发出通知。派生类重写了 on_tick 方法，该方法在启动时调用一次，用于设置目标。让我们看一下继承了 BTActionNode 类的移动节点的实现：

```cpp
include/br2_bt_patrolling/Move.hpp
class Move : public br2_bt_patrolling::BTActionNode<nav2_msgs::action::NavigateToPose>
{
public:
  explicit Move(
    const std::string & xml_tag_name,
    const std::string & action_name,
```

```
include/br2_bt_patrolling/Move.hpp
    const BT::NodeConfiguration & conf);

  void on_tick() override;
  BT::NodeStatus on_success() override;

  static BT::PortsList providedPorts()
  {
    return {
      BT::InputPort<geometry_msgs::msg::PoseStamped>("goal")
    };
  }
};
```

```
src/br2_bt_patrolling/Move.cpp
Move::Move(
  const std::string & xml_tag_name,
  const std::string & action_name,
  const BT::NodeConfiguration & conf)
: br2_bt_patrolling::BtActionNode<nav2_msgs::action::NavigateToPose>(xml_tag_name,
    action_name, conf)
{
}

void
Move::on_tick()
{
  geometry_msgs::msg::PoseStamped goal;
  getInput("goal", goal);

  goal_.pose = goal;
}

BT::NodeStatus
Move::on_success()
{
  RCLCPP_INFO(node_->get_logger(), "navigation Suceeded");

  return BT::NodeStatus::SUCCESS;
}
#include "behaviortree_cpp_v3/bt_factory.h"
BT_REGISTER_NODES(factory)
{
  BT::NodeBuilder builder =
    [](const std::string & name, const BT::NodeConfiguration & config)
    {
      return std::make_unique<br2_bt_patrolling::Move>(
        name, "navigate_to_pose", config);
    };

  factory.registerBuilder<br2_bt_patrolling::Move>(
    "Move", builder);
}
```

- 行为树移动（Move）节点使用 on_success 方法来报告导航已经完成。
- on_tick 方法从输入端口获取目标并将其分配给 goal_ 变量。这个变量将直接被发送到 Nav2。
- 在构建 BTActionNode 类的行为树节点时，第二个参数是 ROS2 动作的名称。在本例中，其值为 navigate_to_pose。

BTLifecycleCtrlNode 是一个类，从它派生出来的行为树节点可以激活/停用生命周期节点。它的创建需要指定要控制的节点的名称。例如在 HeadTracker 的情况下，

它将是 /head_tracker。所有生命周期节点都有不同服务需要管理。在这里，我们将关注两个服务：

- [node name]/get_state：返回生命周期节点（LifeCycleNode）的状态。
- [node name]/set_state：设置生命周期节点（LifeCycleNode）的状态。

让我们看一下 BTLifecycleCtrlNode 实现的代码片段：

```cpp
// include/br2_bt_patrolling/ctrl_support/BTLifecycleCtrlNode.hpp
class BtLifecycleCtrlNode : public BT::ActionNodeBase
{
public:
  BtLifecycleCtrlNode(...)
  : BT::ActionNodeBase(xml_tag_name, conf), ctrl_node_name_(node_name)
  {
  }

  template<typename serviceT>
  typename rclcpp::Client<serviceT>::SharedPtr createServiceClient(
    const std::string & service_name)
  {
    auto srv = node_->create_client<serviceT>(service_name);
    while (!srv->wait_for_service(1s)) {
      ...
    }
    return srv;
  }

  BT::NodeStatus tick() override
  {
    if (status() == BT::NodeStatus::IDLE) {
      change_state_client_ = createServiceClient<lifecycle_msgs::srv::ChangeState>(
        ctrl_node_name_ + "/change_state");
      get_state_client_ = createServiceClient<lifecycle_msgs::srv::GetState>(
        ctrl_node_name_ + "/get_state");
    }

    if (ctrl_node_state_ != lifecycle_msgs::msg::State::PRIMARY_STATE_ACTIVE) {
      ctrl_node_state_ = get_state();
      set_state(lifecycle_msgs::msg::State::PRIMARY_STATE_ACTIVE);
    }

    return BT::NodeStatus::RUNNING;
  }

  void halt() override
  {
    if (ctrl_node_state_ == lifecycle_msgs::msg::State::PRIMARY_STATE_ACTIVE) {
      set_state(lifecycle_msgs::msg::State::PRIMARY_STATE_INACTIVE);
    }
  }

  // Get the state of the controlled node
  uint8_t get_state(){...}

  // Set the state of the controlled node. It can fail, if no transition is possible
  bool set_state(uint8_t state) {...}

  std::string ctrl_node_name_;
  uint8_t ctrl_node_state_;
};
```

在这个类中，两个客户端被实例化：一个用于查询状态，另一个用于设置状态。这两个客户端将分别在 get_state 和 set_state 方法中使用。当节点首次进行执行操作时，请求受控节点切换到活动状态。在停止操作时，请求将其停用。

行为树节点跟踪物体（TrackObjects）节点只需要继承这个类并指定节点的名称：

```
include/br2_bt_patrolling/TrackObjects.hpp
```

```cpp
class TrackObjects : public br2_bt_patrolling::BtLifecycleCtrlNode
{
public:
  explicit TrackObjects(
    const std::string & xml_tag_name,
    const std::string & node_name,
    const BT::NodeConfiguration & conf);

  static BT::PortsList providedPorts()
  {
    return BT::PortsList({});
  }
};
```

```
src/br2_bt_patrolling/TrackObjects.cpp
```

```cpp
TrackObjects::TrackObjects(...)
: br2_bt_patrolling::BtLifecycleCtrlNode(xml_tag_name, action_name, conf)
{
}

#include "behaviortree_cpp_v3/bt_factory.h"
BT_REGISTER_NODES(factory)
{
  BT::NodeBuilder builder =
    [](const std::string & name, const BT::NodeConfiguration & config)
    {
      return std::make_unique<br2_bt_patrolling::TrackObjects>(
        name, "/head_tracker", config);
    };

  factory.registerBuilder<br2_bt_patrolling::TrackObjects>(
    "TrackObjects", builder);
}
```

请注意，跟踪物体（TrackObjects）节点始终返回运行中状态，这就是我们将其作为并行控制节点的子节点的原因。

下面看看其他行为树节点功能是如何实现的：

- **电池检查器节点（BatteryChecker）**：与其他节点的第一个区别是它是一个条件节点。它没有 halt 方法，也不可以返回运行中状态。该节点在每次执行时会检查存储在黑板上的电池电量。如果电池电量低于某个水平，它就返回失败状态。

```
src/br2_bt_patrolling/BatteryChecker.cpp
```

```cpp
const float MIN_LEVEL = 10.0;

BT::NodeStatus
BatteryChecker::tick()
{
  update_battery();

  float battery_level;
  config().blackboard->get("battery_level", battery_level);
  if (battery_level < MIN_LEVEL) {
    return BT::NodeStatus::FAILURE;
  } else {
    return BT::NodeStatus::SUCCESS;
  }
}
```

update_battery 方法从黑板中获取电池电量，并根据时间和当前请求的总速度（last_twist_）减少电量。这只是模拟电池的消耗。

src/br2_bt_patrolling/BatteryChecker.cpp
```cpp
const float DECAY_LEVEL = 0.5;   // 0.5 * |vel| * dt
const float EPSILON = 0.01;      // 0.01 * dt

void
BatteryChecker::update_battery()
{
  float battery_level;
  if (!config().blackboard->get("battery_level", battery_level)) {
    battery_level = 100.0f;
  }
  float dt = (node_->now() - last_reading_time_).seconds();
  last_reading_time_ = node_->now();

  float vel = sqrt(last_twist_.linear.x * last_twist_.linear.x +
    last_twist_.angular.z * last_twist_.angular.z);
  battery_level = std::max(
    0.0f, battery_level - (vel * dt * DECAY_LEVEL) - EPSILON * dt);
  config().blackboard->set("battery_level", battery_level);
}
```

使用 std::max 和 std::min 控制某些计算的范围是有用的。本例中，我们确保电池电量（battery_level）永远不会为负值。

- **充电节点（Recharge）**：它与前一个节点相关。它需要一些时间来充电。请注意，使用黑板可以让一些节点协作更新和测试某些值。

src/br2_bt_patrolling/BatteryChecker.cpp
```cpp
BT::NodeStatus
Recharge::tick()
{
  if (counter_++ < 50) {
    return BT::NodeStatus::RUNNING;
  } else {
    counter_ = 0;
    config().blackboard->set<float>("battery_level", 100.0f);
    return BT::NodeStatus::SUCCESS;
  }
}
```

每个行为树节点都是同一个类的不同实例，但即使某个行为树节点返回了一次成功状态，它仍然准备好了接受下一次触发。在这种情况下，counter_ 计数器变量会重置为 0，开始新的计数。

- **巡逻（Patrol）节点**：它只是让机器人旋转 15s。此节点唯一有趣的地方是它如何控制从第一次执行开始直到返回成功状态的执行时间。请注意，由于节点在第一次执行时是空闲（IDLE）的状态，因此可以存储这一刻的时间戳，以便追踪节点的状态和时间相关的任务。

src/br2_bt_patrolling/Patrol.cpp
```cpp
BT::NodeStatus
Patrol::tick()
{
```

src/br2_bt_patrolling/Patrol.cpp
```cpp
  if (status() == BT::NodeStatus::IDLE) {
    start_time_ = node_->now();
  }

  geometry_msgs::msg::Twist vel_msgs;
  vel_msgs.angular.z = 0.5;
  vel_pub_->publish(vel_msgs);

  auto elapsed = node_->now() - start_time_;

  if (elapsed < 15s) {
    return BT::NodeStatus::RUNNING;
  } else {
    return BT::NodeStatus::SUCCESS;
  }
}
```

- **获取路标（GetWaypoint）节点：** 它存储路标坐标。如果输入端口 wp_id 是字符串"recharge"，则其输出是一个位于地图（map）坐标系中对应于机器人充电器的位置坐标。在其他情况下，每次它被执行都会返回一个不同的路标坐标。

src/br2_bt_patrolling/GetWaypoint.cpp
```cpp
GetWaypoint::GetWaypoint(...)
{
  geometry_msgs::msg::PoseStamped wp;
  wp.header.frame_id = "map";
  wp.pose.orientation.w = 1.0;

  // recharge wp
  wp.pose.position.x = 3.67;
  wp.pose.position.y = -0.24;
  recharge_point_ = wp;

  // wp1
  wp.pose.position.x = 1.07;
  wp.pose.position.y = -12.38;
  waypoints_.push_back(wp);

  // wp2
  wp.pose.position.x = -5.32;
  wp.pose.position.y = -8.85;
  waypoints_.push_back(wp);
}
```

src/br2_bt_patrolling/GetWaypoint.cpp
```cpp
BT::NodeStatus
GetWaypoint::tick()
{
  std::string id;
  getInput("wp_id", id);

  if (id == "recharge") {
    setOutput("waypoint", recharge_point_);
  } else {
    setOutput("waypoint", waypoints_[current_++]);
    current_ = current_ % waypoints_.size();
  }

  return BT::NodeStatus::SUCCESS;
}
```

6.3.5 运行巡逻任务

从实现的角度来看,唯一需要关注的是我们将使用一个启动器来启动主动视觉系统和巡逻节点。值得注意的是,导航和模拟器也可以包含在这个启动器中,但由于它们在屏幕上生成了大量输出信息,因此我们可以在其他终端上手动运行它们。启动器的内容如下:

```python
# br2_navigation/launch/patrolling_launch.py
def generate_launch_description():
    tracking_dir = get_package_share_directory('br2_tracking')

    tracking_cmd = IncludeLaunchDescription(
        PythonLaunchDescriptionSource(os.path.join(tracking_dir, 'launch',
                                                   'tracking.launch.py')))

    patrolling_cmd = Node(
        package='br2_bt_patrolling',
        executable='patrolling_main',
        parameters=[{
          'use_sim_time': True
        }],
        remappings=[
          ('input_scan', '/scan_raw'),
          ('output_vel', '/nav_vel')
        ],
        output='screen'
    )

    ld = LaunchDescription()
    ld.add_action(tracking_cmd)
    ld.add_action(patrolling_cmd)

    return ld
```

因此,请输入下面这些命令,每个命令在一个独立的终端中执行:

```
$ ros2 launch br2_tiago sim.launch.py
```

```
$ ros2 launch br2_navigation tiago_navigation.launch.py
```

Nav2 也使用行为树,并启动了一个使用 Groot 调试其操作的服务器(Groot 是一个图形化工具,用于可视化和调试 ROS2 系统中的行为树)。由于启动时会占用 Groot 的默认端口(1666 和 1667),因此,我们选择在 2666 和 2667 端口上启动它。如果将它们放在同一个端口上,则程序不能正确执行。在连接到巡逻行为树之前,请将端口设置为 2666 和 2667。

```cpp
// src/patrolling_main.cpp
BT::Tree tree = factory.createTreeFromFile(xml_file, blackboard);
auto publisher_zmq = std::make_shared<BT::PublisherZMQ>(tree, 10, 2666, 2667);
```

此时可以选择打开 RViz2 以监视导航,或者打开 Groot 以监视行为树的执行。对于后者,请等待启动巡逻程序以连接到行为树。

```
$ rviz2 --ros-args -p use_sim_time:=true
```

尝试发送一个导航位置，以确保导航正常启动。

```
$ ros2 run groot Groot
```

最后，启动巡逻程序和主动视觉系统：

```
$ ros2 launch br2_bt_patrolling patrolling.launch.py
```

如果一切顺利，那么在电池充电完成后，机器人会在巡逻环境中设立的三个路标。在巡逻过程中，观察机器人如何跟踪检测到的物体。当机器人到达一个路标并转身时，注意跟踪活动是如何停止的。经过一段时间的操作，机器人将再次耗尽电量，再一次前往充电地点，然后再继续巡逻。

建议的练习

1）用行为树创建一个程序，使机器人持续移动到没有障碍物的空间。
2）探索 Nav2 的功能：
- 在每个机器人不应进入的房间中心标记禁止区域。
- 修改行为树导航器内部的行为树，使导航总是在目标前 1m 结束。
- 尝试不同的控制器/规划器算法。

3）在巡逻时将检测到的对象作为 3D 边界框⊖，并将其发布。你可以通过以下方式实现：
- 使用点云。
- 使用深度图像和 CameraInfo 信息。

⊖ 3D 边界框（3D bounding box）是一个立方体或矩形，用于包围一个三维对象或场景中的物体，在计算机视觉和机器人领域中，3D 边界框常用于表示和处理检测到的物体的位置和形状。——译者注

APPENDIX

附录

源代码

A.1 BR2_BASICS 软件包

```
Package br2_basics

br2_basics
├── CMakeLists.txt
├── config
│   └── params.yaml
├── launch
│   ├── includer_launch.py
│   ├── param_node_v1_launch.py
│   ├── param_node_v2_launch.py
│   ├── pub_sub_v1_launch.py
│   └── pub_sub_v2_launch.py
├── package.xml
└── src
    ├── executors.cpp
    ├── logger_class.cpp
    ├── logger.cpp
    ├── param_reader.cpp
    ├── publisher_class.cpp
    ├── publisher.cpp
    └── subscriber_class.cpp
```

```
br2_basics/CMakeLists.txt

cmake_minimum_required(VERSION 3.5)
project(br2_basics)

find_package(ament_cmake REQUIRED)
find_package(rclcpp REQUIRED)
find_package(std_msgs REQUIRED)

set(dependencies
    rclcpp
    std_msgs
)
```

`br2_basics/CMakeLists.txt`

```cmake
add_executable(publisher src/publisher.cpp)
ament_target_dependencies(publisher ${dependencies})

add_executable(publisher_class src/publisher_class.cpp)
ament_target_dependencies(publisher_class ${dependencies})

add_executable(subscriber_class src/subscriber_class.cpp)
ament_target_dependencies(subscriber_class ${dependencies})

add_executable(executors src/executors.cpp)
ament_target_dependencies(executors ${dependencies})

add_executable(logger src/logger.cpp)
ament_target_dependencies(logger ${dependencies})

add_executable(logger_class src/logger_class.cpp)
ament_target_dependencies(logger_class ${dependencies})

add_executable(param_reader src/param_reader.cpp)
ament_target_dependencies(param_reader ${dependencies})

install(TARGETS
  publisher
  publisher_class
  subscriber_class
  executors
  logger
  logger_class
  param_reader
  ARCHIVE DESTINATION lib
  LIBRARY DESTINATION lib
  RUNTIME DESTINATION lib/${PROJECT_NAME}
)

install(DIRECTORY launch config DESTINATION share/${PROJECT_NAME})

if(BUILD_TESTING)
  find_package(ament_lint_auto REQUIRED)
  ament_lint_auto_find_test_dependencies()
endif()

ament_export_dependencies(${dependencies})
ament_package()
```

`br2_basics/launch/param_node_v2_launch.py`

```python
import os

from ament_index_python.packages import get_package_share_directory

from launch import LaunchDescription
from launch_ros.actions import Node

def generate_launch_description():

  pkg_dir = get_package_share_directory('basics')
  param_file = os.path.join(pkg_dir, 'config', 'params.yaml')

  param_reader_cmd = Node(
    package='br2_basics',
    executable='param_reader',
    parameters=[param_file],
    output='screen'
  )

  ld = LaunchDescription()
  ld.add_action(param_reader_cmd)

  return ld
```

br2_basics/launch/pub_sub_v2_launch.py

```python
from launch import LaunchDescription
from launch_ros.actions import Node

def generate_launch_description():
    return LaunchDescription([
        Node(
            package='br2_basics',
            executable='publisher',
            output='screen'
        ),
        Node(
            package='br2_basics',
            executable='subscriber_class',
            output='screen'
        )
    ])
```

br2_basics/launch/pub_sub_v1_launch.py

```python
from launch import LaunchDescription
from launch_ros.actions import Node

def generate_launch_description():
    pub_cmd = Node(
        package='br2_basics',
        executable='publisher',
        output='screen'
    )
    sub_cmd = Node(
        package='br2_basics',
        executable='subscriber_class',
        output='screen'
    )

    ld = LaunchDescription()
    ld.add_action(pub_cmd)
    ld.add_action(sub_cmd)

    return ld
```

br2_basics/launch/includer_launch.py

```python
import os

from ament_index_python.packages import get_package_share_directory
from launch import LaunchDescription
from launch.actions import IncludeLaunchDescription
from launch.launch_description_sources import PythonLaunchDescriptionSource

def generate_launch_description():
    return LaunchDescription([
        IncludeLaunchDescription(
            PythonLaunchDescriptionSource(os.path.join(
                get_package_share_directory('br2_basics'),
                'launch',
                'pub_sub_v2_launch.py'))
        )
    ])
```

br2_basics/launch/param_node_v1_launch.py

```python
from launch import LaunchDescription
from launch_ros.actions import Node
```

```
br2_basics/launch/param_node_v1_launch.py
```
```python
def generate_launch_description():

    param_reader_cmd = Node(
        package='br2_basics',
        executable='param_reader',
        parameters=[{
          'particles': 300,
          'topics': ['scan', 'image'],
          'topic_types': ['sensor_msgs/msg/LaserScan', 'sensor_msgs/msg/Image']
        }],
        output='screen'
    )

    ld = LaunchDescription()
    ld.add_action(param_reader_cmd)

    return ld
```

```
br2_basics/package.xml
```
```xml
<?xml version="1.0"?>
<?xml-model href="http://download.ros.org/schema/package_format3.xsd"
  schematypens="http://www.w3.org/2001/XMLSchema"?>
<package format="3">
  <name>br2_basics</name>
  <version>0.0.0</version>
  <description>Basic nodes for ROS 2 introduction</description>
  <maintainer email="fmrico@gmail.com">Francisco Martín</maintainer>
  <license>Apache 2</license>

  <buildtool_depend>ament_cmake</buildtool_depend>

  <depend>rclcpp</depend>
  <depend>std_msgs</depend>

  <test_depend>ament_lint_auto</test_depend>
  <test_depend>ament_lint_common</test_depend>

  <export>
    <build_type>ament_cmake</build_type>
  </export>
</package>
```

```
br2_basics/config/params.yaml
```
```yaml
localization_node:
  ros__parameters:
    number_particles: 300
    topics: [scan, image]
    topic_types: [sensor_msgs/msg/LaserScan, sensor_msgs/msg/Image]
```

```
br2_basics/src/logger_class.cpp
```
```cpp
#include "rclcpp/rclcpp.hpp"

using namespace std::chrono_literals;

class LoggerNode : public rclcpp::Node
{
public:
  LoggerNode() : Node("logger_node")
  {
    counter_ = 0;
    timer_ = create_wall_timer(
      500ms, std::bind(&LoggerNode::timer_callback, this));
  }
```

br2_basics/src/logger_class.cpp

```cpp
  void timer_callback()
  {
    RCLCPP_INFO(get_logger(), "Hello %d", counter_++);
  }
private:
  rclcpp::TimerBase::SharedPtr timer_;
  int counter_;
};

int main(int argc, char * argv[]) {
  rclcpp::init(argc, argv);

  auto node = std::make_shared<LoggerNode>();

  rclcpp::spin(node);

  rclcpp::shutdown();
  return 0;
```

br2_basics/src/subscriber_class.cpp

```cpp
#include "rclcpp/rclcpp.hpp"
#include "std_msgs/msg/int32.hpp"

using std::placeholders::_1;

class SubscriberNode : public rclcpp::Node
{
public:
  SubscriberNode() : Node("subscriber_node")
  {
    subscriber_ = create_subscription<std_msgs::msg::Int32>("int_topic", 10,
      std::bind(&SubscriberNode::callback, this, _1));
  }

  void callback(const std_msgs::msg::Int32::SharedPtr msg)
  {
    RCLCPP_INFO(get_logger(), "Hello %d", msg->data);
  }
private:
  rclcpp::Subscription<std_msgs::msg::Int32>::SharedPtr subscriber_;
};

int main(int argc, char * argv[]) {
  rclcpp::init(argc, argv);

  auto node = std::make_shared<SubscriberNode>();

  rclcpp::spin(node);

  rclcpp::shutdown();
  return 0;
```

br2_basics/src/publisher.cpp

```cpp
#include "rclcpp/rclcpp.hpp"
#include "std_msgs/msg/int32.hpp"

using namespace std::chrono_literals;

int main(int argc, char * argv[]) {
  rclcpp::init(argc, argv);

  auto node = rclcpp::Node::make_shared("publisher_node");
  auto publisher = node->create_publisher<std_msgs::msg::Int32>(
    "int_topic", 10);
```

br2_basics/src/publisher.cpp

```cpp
  std_msgs::msg::Int32 message;
  message.data = 0;

  rclcpp:Rate loop_rate(500ms);
  while (rclcpp::ok()) {
    message.data += 1;
    publisher->publish(message);

    rclcpp::spin_some(node);
    loop_rate.sleep();
  }

  rclcpp::shutdown();
  return 0;
}
```

br2_basics/src/param_reader.cpp

```cpp
#include <vector>
#include <string>

#include "rclcpp/rclcpp.hpp"

class LocalizationNode : public rclcpp::Node
{
public:
  LocalizationNode() : Node("localization_node")
  {
    declare_parameter("number_particles", 200);
    declare_parameter("topics", std::vector<std::string>());
    declare_parameter("topic_types", std::vector<std::string>());

    get_parameter("number_particles", num_particles_);
    RCLCPP_INFO_STREAM(get_logger(), "Number of particles: " << num_particles_);

    get_parameter("topics", topics_);
    get_parameter("topic_types", topic_types_);

    if (topics_.size() != topic_types_.size()) {
      RCLCPP_ERROR(get_logger(), "Number of topics (%zu) != number of types (%zu)",
        topics_.size(), topic_types_.size());
    } else {
      RCLCPP_INFO_STREAM(get_logger(), "Number of topics: " << topics_.size());
      for (size_t i = 0; i < topics_.size(); i++) {
        RCLCPP_INFO_STREAM(get_logger(), "\t" << topics_[i] << "\t - " << topic_types_[i]);
      }
    }
  }

private:
  int num_particles_;
  std::vector<std::string> topics_;
  std::vector<std::string> topic_types_;
};

int main(int argc, char * argv[]) {
  rclcpp::init(argc, argv);

  auto node = std::make_shared<LocalizationNode>();

  rclcpp::spin(node);

  rclcpp::shutdown();
  return 0;
}
```

br2_basics/src/executors.cpp

```cpp
#include "rclcpp/rclcpp.hpp"

#include "std_msgs/msg/int32.hpp"
```

br2_basics/src/executors.cpp

```cpp
using namespace std::chrono_literals;
using std::placeholders::_1;

class PublisherNode : public rclcpp::Node
{
public:
  PublisherNode() : Node("publisher_node")
  {
    publisher_ = create_publisher<std_msgs::msg::Int32>("int_topic", 10);
    timer_ = create_wall_timer(
      500ms, std::bind(&PublisherNode::timer_callback, this));
  }

  void timer_callback()
  {
    message_.data += 1;
    publisher_->publish(message_);
  }
private:
  rclcpp::Publisher<std_msgs::msg::Int32>::SharedPtr publisher_;
  rclcpp::TimerBase::SharedPtr timer_;
  std_msgs::msg::Int32 message_;
};

class SubscriberNode : public rclcpp::Node
{
public:
  SubscriberNode() : Node("subscriber_node")
  {
    subscriber_ = create_subscription<std_msgs::msg::Int32>("int_topic", 10,
      std::bind(&SubscriberNode::callback, this, _1));
  }

  void callback(const std_msgs::msg::Int32::SharedPtr msg)
  {
    RCLCPP_INFO(get_logger(), "Hello %d", msg->data);
  }
private:
  rclcpp::Subscription<std_msgs::msg::Int32>::SharedPtr subscriber_;
};

int main(int argc, char * argv[]) {
  rclcpp::init(argc, argv);

  auto node_pub = std::make_shared<PublisherNode>();
  auto node_sub = std::make_shared<SubscriberNode>();

  rclcpp::executors::SingleThreadedExecutor executor;
  // rclcpp::executors::MultiThreadedExecutor executor(
  //    rclcpp::executor::ExecutorArgs(), 8);

  executor.add_node(node_pub);
  executor.add_node(node_sub);

  executor.spin();

  rclcpp::shutdown();
  return 0;
}
```

br2_basics/src/publisher_class.cpp

```cpp
#include "rclcpp/rclcpp.hpp"

#include "std_msgs/msg/int32.hpp"

using namespace std::chrono_literals;
using std::placeholders::_1;
```

br2_basics/src/publisher_class.cpp

```cpp
class PublisherNode : public rclcpp::Node
{
public:
  PublisherNode() : Node("publisher_node")
  {
    publisher_ = create_publisher<std_msgs::msg::Int32>("int_topic", 10);
    timer_ = create_wall_timer(
      500ms, std::bind(&PublisherNode::timer_callback, this));
  }

  void timer_callback()
  {
    message_.data += 1;
    publisher_->publish(message_);
  }
private:
  rclcpp::Publisher<std_msgs::msg::Int32>::SharedPtr publisher_;
  rclcpp::TimerBase::SharedPtr timer_;
  std_msgs::msg::Int32 message_;
};

int main(int argc, char * argv[]) {
  rclcpp::init(argc, argv);

  auto node = std::make_shared<PublisherNode>();

  rclcpp::spin(node);

  rclcpp::shutdown();
  return 0;
}
```

br2_basics/src/logger.cpp

```cpp
#include "rclcpp/rclcpp.hpp"

using namespace std::chrono_literals;

int main(int argc, char * argv[]) {
  rclcpp::init(argc, argv);

  auto node = rclcpp::Node::make_shared("logger_node");

  rclcpp::Rate loop_rate(500ms);
  int counter = 0;
  while (rclcpp::ok()) {
    RCLCPP_INFO(node->get_logger(), "Hello %d", counter++);

    rclcpp::spin_some(node);
    loop_rate.sleep();
  }

  rclcpp::shutdown();
  return 0;
}
```

A.2 BR2_FSM_BUMPGO_CPP 软件包

Package br2_fsm_bumpgo_cpp

```
br2_fsm_bumpgo_cpp
├── CMakeLists.txt
├── include
│   └── br2_fsm_bumpgo_cpp
│       └── BumpGoNode.hpp
```

Package br2_fsm_bumpgo_cpp

```
├── launch
│   └── bump_and_go.launch.py
├── package.xml
└── src
    ├── br2_fsm_bumpgo_cpp
    │   └── BumpGoNode.cpp
```

br2_fsm_bumpgo_cpp/CMakeLists.txt

```cmake
cmake_minimum_required(VERSION 3.5)
project(br2_fsm_bumpgo_cpp)

set(CMAKE_CXX_STANDARD 17)

find_package(ament_cmake REQUIRED)
find_package(rclcpp REQUIRED)
find_package(sensor_msgs REQUIRED)
find_package(geometry_msgs REQUIRED)

set(dependencies
  rclcpp
  sensor_msgs
  geometry_msgs
)

include_directories(include)

add_executable(bumpgo
  src/br2_fsm_bumpgo_cpp/BumpGoNode.cpp
  src/bumpgo_main.cpp
)
ament_target_dependencies(bumpgo ${dependencies})

install(TARGETS
  bumpgo
  ARCHIVE DESTINATION lib
  LIBRARY DESTINATION lib
  RUNTIME DESTINATION lib/${PROJECT_NAME}
)

install(DIRECTORY launch DESTINATION share/${PROJECT_NAME})

if(BUILD_TESTING)
  find_package(ament_lint_auto REQUIRED)
  ament_lint_auto_find_test_dependencies()

  set(ament_cmake_cpplint_FOUND TRUE)
  ament_lint_auto_find_test_dependencies()
endif()

ament_package()
```

br2_fsm_bumpgo_cpp/launch/bump_and_go.launch.py

```python
from launch import LaunchDescription
from launch_ros.actions import Node


def generate_launch_description():

    bumpgo_cmd = Node(package='br2_fsm_bumpgo_cpp',
                      executable='bumpgo',
                      output='screen',
                      parameters=[{
                          'use_sim_time': True
                      }],
                      remappings=[
                          ('input_scan', '/scan_raw'),
                          ('output_vel', '/nav_vel')
                      ])
```

br2_fsm_bumpgo_cpp/launch/bump_and_go.launch.py

```python
    ld = LaunchDescription()
    ld.add_action(bumpgo_cmd)

    return ld
```

br2_fsm_bumpgo_cpp/package.xml

```xml
<?xml version="1.0"?>
<?xml-model href="http://download.ros.org/schema/package_format3.xsd"
    schematypens="http://www.w3.org/2001/XMLSchema"?>
<package format="3">
  <name>br2_fsm_bumpgo_cpp</name>
  <version>0.1.0</version>
  <description>Bump and Go behavior based on Finite State Machines</description>
  <maintainer email="fmrico@gmail.com">Francisco Martín</maintainer>
  <license>Apache 2.0</license>

  <buildtool_depend>ament_cmake</buildtool_depend>

  <depend>rclcpp</depend>
  <depend>geometry_msgs</depend>
  <depend>sensor_msgs</depend>

  <test_depend>ament_lint_auto</test_depend>
  <test_depend>ament_lint_common</test_depend>

  <export>
    <build_type>ament_cmake</build_type>
  </export>
</package>
```

br2_fsm_bumpgo_cpp/include/br2_fsm_bumpgo_cpp/BumpGoNode.hpp

```cpp
#ifndef BR2_BT_BUMPGO__BUMPGONODE_HPP_
#define BR2_BT_BUMPGO__BUMPGONODE_HPP_

#include "sensor_msgs/msg/laser_scan.hpp"
#include "geometry_msgs/msg/twist.hpp"

#include "rclcpp/rclcpp.hpp"

namespace br2_fsm_bumpgo_cpp
{

using namespace std::chrono_literals;

class BumpGoNode : public rclcpp::Node
{
public:
  BumpGoNode();

private:
  void scan_callback(sensor_msgs::msg::LaserScan::UniquePtr msg);
  void control_cycle();

  static const int FORWARD = 0;
  static const int BACK = 1;
  static const int TURN = 2;
  static const int STOP = 3;
  int state_;
  rclcpp::Time state_ts_;

  void go_state(int new_state);
  bool check_forward_2_back();
  bool check_forward_2_stop();
  bool check_back_2_turn();
  bool check_turn_2_forward();
  bool check_stop_2_forward();
```

br2_fsm_bumpgo_cpp/include/br2_fsm_bumpgo_cpp/BumpGoNode.hpp

```cpp
  const rclcpp::Duration TURNING_TIME {2s};
  const rclcpp::Duration BACKING_TIME {2s};
  const rclcpp::Duration SCAN_TIMEOUT {1s};

  static constexpr float SPEED_LINEAR = 0.3f;
  static constexpr float SPEED_ANGULAR = 0.3f;
  static constexpr float OBSTACLE_DISTANCE = 1.0f;

  rclcpp::Subscription<sensor_msgs::msg::LaserScan>::SharedPtr scan_sub_;
  rclcpp::Publisher<geometry_msgs::msg::Twist>::SharedPtr vel_pub_;
  rclcpp::TimerBase::SharedPtr timer_;

  sensor_msgs::msg::LaserScan::UniquePtr last_scan_;
};

}  // namespace br2_fsm_bumpgo_cpp

#endif  // BR2_BT_BUMPGO__BUMPGONODE_HPP_
```

br2_fsm_bumpgo_cpp/src/bumpgo_main.cpp

```cpp
#include <memory>

#include "br2_fsm_bumpgo_cpp/BumpGoNode.hpp"
#include "rclcpp/rclcpp.hpp"

int main(int argc, char * argv[])
{
  rclcpp::init(argc, argv);

  auto bumpgo_node = std::make_shared<br2_fsm_bumpgo_cpp::BumpGoNode>();
  rclcpp::spin(bumpgo_node);

  rclcpp::shutdown();

  return 0;
}
```

br2_fsm_bumpgo_cpp/src/br2_fsm_bumpgo_cpp/BumpGoNode.cpp

```cpp
#include <utility>
#include "br2_fsm_bumpgo_cpp/BumpGoNode.hpp"

#include "sensor_msgs/msg/laser_scan.hpp"
#include "geometry_msgs/msg/twist.hpp"

#include "rclcpp/rclcpp.hpp"

namespace br2_fsm_bumpgo_cpp
{

using namespace std::chrono_literals;
using std::placeholders::_1;

BumpGoNode::BumpGoNode()
: Node("bump_go"),
  state_(FORWARD)
{
  scan_sub_ = create_subscription<sensor_msgs::msg::LaserScan>(
    "input_scan", rclcpp::SensorDataQoS(),
    std::bind(&BumpGoNode::scan_callback, this, _1));

  vel_pub_ = create_publisher<geometry_msgs::msg::Twist>("output_vel", 10);
  timer_ = create_wall_timer(50ms, std::bind(&BumpGoNode::control_cycle, this));

  state_ts_ = now();
}
```

```
br2_fsm_bumpgo_cpp/src/br2_fsm_bumpgo_cpp/BumpGoNode.cpp
void
BumpGoNode::scan_callback(sensor_msgs::msg::LaserScan::UniquePtr msg)
{
  last_scan_ = std::move(msg);
}

void
BumpGoNode::control_cycle()
{
  // Do nothing until the first sensor read
  if (last_scan_ == nullptr)
    return;

  geometry_msgs::msg::Twist out_vel;

  switch (state_) {
    case FORWARD:
      out_vel.linear.x = SPEED_LINEAR;

      if (check_forward_2_stop())
        go_state(STOP);
      if (check_forward_2_back())
        go_state(BACK);

      break;
    case BACK:
      out_vel.linear.x = -SPEED_LINEAR;

      if (check_back_2_turn())
        go_state(TURN);

      break;
    case TURN:
      out_vel.angular.z = SPEED_ANGULAR;

      if (check_turn_2_forward())
        go_state(FORWARD);

      break;
    case STOP:
      if (check_stop_2_forward())
        go_state(FORWARD);
      break;
  }

  vel_pub_->publish(out_vel);
}

void
BumpGoNode::go_state(int new_state)
{
  state_ = new_state;
  state_ts_ = now();
}

bool
BumpGoNode::check_forward_2_back()
{
  // going forward when deteting an obstacle
  // at 0.5 meters with the front laser read
  size_t pos = last_scan_->ranges.size() / 2;
  return last_scan_->ranges[pos] < OBSTACLE_DISTANCE;
}

bool
BumpGoNode::check_forward_2_stop()
{
  // Stop if no sensor readings for 1 second
  auto elapsed = now() - rclcpp::Time(last_scan_->header.stamp);
  return elapsed > SCAN_TIMEOUT;
}
```

```
br2_fsm_bumpgo_cpp/src/br2_fsm_bumpgo_cpp/BumpGoNode.cpp
```
```cpp
bool
BumpGoNode::check_stop_2_forward()
{
  // Going forward if sensor readings are available
  // again
  auto elapsed = now() - rclcpp::Time(last_scan_->header.stamp);
  return elapsed < SCAN_TIMEOUT;
}

bool
BumpGoNode::check_back_2_turn()
{
  // Going back for 2 seconds
  return (now() - state_ts_) > BACKING_TIME;
}

bool
BumpGoNode::check_turn_2_forward()
{
  // Turning for 2 seconds
  return (now() - state_ts_) > TURNING_TIME;
}
```

A.3　BR2_FSM_BUMPGO_PY 软件包

```
Package br2_fsm_bumpgo_py
```
```
br2_fsm_bumpgo_py
├── br2_fsm_bumpgo_py
│   ├── bump_go_main.py
│   └── __init__.py
├── launch
│   └── bump_and_go.launch.py
├── package.xml
├── resource
│   └── br2_fsm_bumpgo_py
├── setup.cfg
├── setup.py
└── test
    ├── test_copyright.py
    ├── test_flake8.py
    └── test_pep257.py
```

```
br2_fsm_bumpgo_py/launch/bump_and_go.launch.py
```
```python
from launch import LaunchDescription
from launch_ros.actions import Node


def generate_launch_description():

    kobuki_cmd = Node(package='br2_fsm_bumpgo_py',
                executable='bump_go_main',
                output='screen',
                parameters=[{
                  'use_sim_time': True
                }],
                remappings=[
                  ('input_scan', '/scan_raw'),
                  ('output_vel', '/nav_vel')
                ])

    ld = LaunchDescription()
    ld.add_action(kobuki_cmd)

    return ld
```

br2_fsm_bumpgo_py/package.xml

```xml
<?xml version="1.0"?>
<?xml-model href="http://download.ros.org/schema/package_format3.xsd"
  schematypens="http://www.w3.org/2001/XMLSchema"?>
<package format="3">
  <name>br2_fsm_bumpgo_py</name>
  <version>0.0.0</version>
  <description>BumpGo behavior based on Finite Satate Machines in Python</description>
  <maintainer email="fmrico@gmail.com">Francisco Martŷn</maintainer>
  <license>Apache 2.0</license>

  <depend>rclcpy</depend>
  <depend>sensor_msgs</depend>
  <depend>geometry_msgs</depend>

  <test_depend>ament_copyright</test_depend>
  <test_depend>ament_flake8</test_depend>
  <test_depend>ament_pep257</test_depend>
  <test_depend>python3-pytest</test_depend>

  <export>
    <build_type>ament_python</build_type>
  </export>
</package>
```

br2_fsm_bumpgo_py/setup.py

```python
import os
from glob import glob

from setuptools import setup

package_name = 'br2_fsm_bumpgo_py'

setup(
    name=package_name,
    version='0.0.0',
    packages=[package_name],
    data_files=[
        ('share/ament_index/resource_index/packages',
            ['resource/' + package_name]),
        ('share/' + package_name, ['package.xml']),
        (os.path.join('share', package_name, 'launch'), glob('launch/*.launch.py'))
    ],
    install_requires=['setuptools'],
    zip_safe=True,
    maintainer='Francisco Martŷn',
    maintainer_email='fmrico@gmail.com',
    description='BumpGo behavior based on Finite Satate Machines in Python',
    license='Apache 2.0',
    tests_require=['pytest'],
    entry_points={
        'console_scripts': [
            'bump_go_main = br2_fsm_bumpgo_py.bump_go_main:main'
        ],
    },
)
```

br2_fsm_bumpgo_py/bump_go_py.py

```python
import rclpy

from rclpy.duration import Duration
from rclpy.node import Node
from rclpy.qos import qos_profile_sensor_data
from rclpy.time import Time

from geometry_msgs.msg import Twist
from sensor_msgs.msg import LaserScan
```

`br2_fsm_bumpgo_py/bump_go_py.py`

```python
class BumpGoNode(Node):
    def __init__(self):
        super().__init__('bump_go')

        self.FORWARD = 0
        self.BACK = 1
        self.TURN = 2
        self.STOP = 3
        self.state = self.FORWARD
        self.state_ts = self.get_clock().now()

        self.TURNING_TIME = 2.0
        self.BACKING_TIME = 2.0
        self.SCAN_TIMEOUT = 1.0

        self.SPEED_LINEAR = 0.3
        self.SPEED_ANGULAR = 0.3
        self.OBSTACLE_DISTANCE = 1.0

        self.last_scan = None

        self.scan_sub = self.create_subscription(
            LaserScan,
            'input_scan',
            self.scan_callback,
            qos_profile_sensor_data)

        self.vel_pub = self.create_publisher(Twist, 'output_vel', 10)
        self.timer = self.create_timer(0.05, self.control_cycle)

    def scan_callback(self, msg):
        self.last_scan = msg

    def control_cycle(self):
        if self.last_scan is None:
            return

        out_vel = Twist()

        if self.state == self.FORWARD:
            out_vel.linear.x = self.SPEED_LINEAR

            if self.check_forward_2_stop():
                self.go_state(self.STOP)
            if self.check_forward_2_back():
                self.go_state(self.BACK)

        elif self.state == self.BACK:
            out_vel.linear.x = -self.SPEED_LINEAR

            if self.check_back_2_turn():
                self.go_state(self.TURN)

        elif self.state == self.TURN:
            out_vel.angular.z = self.SPEED_ANGULAR

            if self.check_turn_2_forward():
                self.go_state(self.FORWARD)

        elif self.state == self.STOP:
            if self.check_stop_2_forward():
                self.go_state(self.FORWARD)

        self.vel_pub.publish(out_vel)

    def go_state(self, new_state):
        self.state = new_state
        self.state_ts = self.get_clock().now()
```

```
br2_fsm_bumpgo_py/setup.py
    def check_forward_2_back(self):
        pos = round(len(self.last_scan.ranges) / 2)
        return self.last_scan.ranges[pos] < self.OBSTACLE_DISTANCE

    def check_forward_2_stop(self):
        elapsed = self.get_clock().now() - Time.from_msg(self.last_scan.header.stamp)
        return elapsed > Duration(seconds=self.SCAN_TIMEOUT)

    def check_stop_2_forward(self):
        elapsed = self.get_clock().now() - Time.from_msg(self.last_scan.header.stamp)
        return elapsed < Duration(seconds=self.SCAN_TIMEOUT)

    def check_back_2_turn(self):
        elapsed = self.get_clock().now() - self.state_ts
        return elapsed > Duration(seconds=self.BACKING_TIME)

    def check_turn_2_forward(self):
        elapsed = self.get_clock().now() - self.state_ts
        return elapsed > Duration(seconds=self.TURNING_TIME)

def main(args=None):
    rclpy.init(args=args)

    bump_go_node = BumpGoNode()

    rclpy.spin(bump_go_node)

    bump_go_node.destroy_node()
    rclpy.shutdown()

if __name__ == '__main__':
    main()
```

A.4 BR2_TF2_DETECTOR 软件包

```
Package br2_tf2_detector
br2_tf2_detector
├── CMakeLists.txt
├── include
│   └── br2_tf2_detector
│       ├── ObstacleDetectorImprovedNode.hpp
│       ├── ObstacleDetectorNode.hpp
│       └── ObstacleMonitorNode.hpp
├── launch
│   ├── detector_basic.launch.py
│   └── detector_improved.launch.py
├── package.xml
└── src
    ├── br2_tf2_detector
    │   ├── ObstacleDetectorImprovedNode.cpp
    │   ├── ObstacleDetectorNode.cpp
    │   └── ObstacleMonitorNode.cpp
    └── detector_improved_main.cpp
```

```
br2_tf2_detector/CMakeLists.txt
cmake_minimum_required(VERSION 3.5)
project(br2_tf2_detector)

set(CMAKE_CXX_STANDARD 17)

# find dependencies
find_package(ament_cmake REQUIRED)
find_package(rclcpp REQUIRED)
find_package(tf2_ros REQUIRED)
```

br2_tf2_detector/CMakeLists.txt

```cmake
find_package(geometry_msgs REQUIRED)
find_package(sensor_msgs REQUIRED)
find_package(visualization_msgs REQUIRED)

set(dependencies
    rclcpp
    tf2_ros
    geometry_msgs
    sensor_msgs
    visualization_msgs
)

include_directories(include)

add_library(${PROJECT_NAME} SHARED
  src/br2_tf2_detector/ObstacleDetectorNode.cpp
  src/br2_tf2_detector/ObstacleMonitorNode.cpp
  src/br2_tf2_detector/ObstacleDetectorImprovedNode.cpp
)
ament_target_dependencies(${PROJECT_NAME} ${dependencies})

add_executable(detector src/detector_main.cpp)
ament_target_dependencies(detector ${dependencies})
target_link_libraries(detector ${PROJECT_NAME})

add_executable(detector_improved src/detector_improved_main.cpp)
ament_target_dependencies(detector_improved ${dependencies})
target_link_libraries(detector_improved ${PROJECT_NAME})

install(TARGETS
  ${PROJECT_NAME}
  detector
  detector_improved
  ARCHIVE DESTINATION lib
  LIBRARY DESTINATION lib
  RUNTIME DESTINATION lib/${PROJECT_NAME}
)

install(DIRECTORY launch DESTINATION share/${PROJECT_NAME})

if(BUILD_TESTING)
  find_package(ament_lint_auto REQUIRED)
  ament_lint_auto_find_test_dependencies()
endif()

ament_package()
```

br2_tf2_detector/launch/detector_improved.launch.py

```python
from launch import LaunchDescription
from launch_ros.actions import Node


def generate_launch_description():

    detector_cmd = Node(package='br2_tf2_detector',
                        executable='detector_improved',
                        output='screen',
                        parameters=[{
                          'use_sim_time': True
                        }],
                        remappings=[
                          ('input_scan', '/scan_raw')
                        ])

    ld = LaunchDescription()
    ld.add_action(detector_cmd)

    return ld
```

br2_tf2_detector/launch/detector_basic.launch.py

```python
from launch import LaunchDescription
from launch_ros.actions import Node


def generate_launch_description():

    detector_cmd = Node(package='br2_tf2_detector',
                       executable='detector',
                       output='screen',
                       parameters=[{
                         'use_sim_time': True
                       }],
                       remappings=[
                         ('input_scan', '/scan_raw')
                       ])

    ld = LaunchDescription()
    ld.add_action(detector_cmd)

    return ld
```

br2_tf2_detector/package.xml

```xml
<?xml version="1.0"?>
<?xml-model href="http://download.ros.org/schema/package_format3.xsd"
  schematypens="http://www.w3.org/2001/XMLSchema"?>
<package format="3">
  <name>br2_tf2_detector</name>
  <version>0.0.0</version>
  <description>TF2 detector package</description>
  <maintainer email="fmrico@gmail.com">Francisco Martín</maintainer>
  <license>Apache 2.0</license>

  <buildtool_depend>ament_cmake</buildtool_depend>

  <depend>rclcpp</depend>
  <depend>tf2_ros</depend>
  <depend>geometry_msgs</depend>
  <depend>sensor_msgs</depend>

  <test_depend>ament_lint_auto</test_depend>
  <test_depend>ament_lint_common</test_depend>

  <export>
    <build_type>ament_cmake</build_type>
  </export>
</package>
```

br2_tf2_detector/include/br2_tf2_detector/ObstacleDetectorImprovedNode.hpp

```cpp
#ifndef BR2_TF2_DETECTOR__OBSTACLEDETECTORIMPROVEDNODE_HPP_
#define BR2_TF2_DETECTOR__OBSTACLEDETECTORIMPROVEDNODE_HPP_

#include <tf2_ros/static_transform_broadcaster.h>
#include <tf2_ros/buffer.h>
#include <tf2_ros/transform_listener.h>

#include <memory>

#include "sensor_msgs/msg/laser_scan.hpp"

#include "rclcpp/rclcpp.hpp"

namespace br2_tf2_detector
{

class ObstacleDetectorImprovedNode : public rclcpp::Node
{
```

br2_tf2_detector/include/br2_tf2_detector/ObstacleDetectorImprovedNode.hpp

```cpp
public:
  ObstacleDetectorImprovedNode();

private:
  void scan_callback(sensor_msgs::msg::LaserScan::UniquePtr msg);

  rclcpp::Subscription<sensor_msgs::msg::LaserScan>::SharedPtr scan_sub_;
  std::shared_ptr<tf2_ros::StaticTransformBroadcaster> tf_broadcaster_;

  tf2::BufferCore tf_buffer_;
  tf2_ros::TransformListener tf_listener_;
};

}  // namespace br2_tf2_detector
#endif  // BR2_TF2_DETECTOR__OBSTACLEDETECTORIMPROVEDNODE_HPP_
```

br2_tf2_detector/include/br2_tf2_detector/ObstacleMonitorNode.hpp

```cpp
#ifndef BR2_TF2_DETECTOR__OBSTACLEMONITORNODE_HPP_
#define BR2_TF2_DETECTOR__OBSTACLEMONITORNODE_HPP_

#include <tf2_ros/buffer.h>
#include <tf2_ros/transform_listener.h>

#include <memory>

#include "sensor_msgs/msg/laser_scan.hpp"
#include "visualization_msgs/msg/marker.hpp"

#include "rclcpp/rclcpp.hpp"

namespace br2_tf2_detector
{

class ObstacleMonitorNode : public rclcpp::Node
{
public:
  ObstacleMonitorNode();

private:
  void control_cycle();
  rclcpp::TimerBase::SharedPtr timer_;

  tf2::BufferCore tf_buffer_;
  tf2_ros::TransformListener tf_listener_;

  rclcpp::Publisher<visualization_msgs::msg::Marker>::SharedPtr marker_pub_;
};

}  // namespace br2_tf2_detector
#endif  // BR2_TF2_DETECTOR__OBSTACLEMONITORNODE_HPP_
```

br2_tf2_detector/include/br2_tf2_detector/ObstacleDetectorNode.hpp

```cpp
#ifndef BR2_TF2_DETECTOR__OBSTACLEDETECTORNODE_HPP_
#define BR2_TF2_DETECTOR__OBSTACLEDETECTORNODE_HPP_

#include <tf2_ros/static_transform_broadcaster.h>

#include <memory>

#include "sensor_msgs/msg/laser_scan.hpp"

#include "rclcpp/rclcpp.hpp"

namespace br2_tf2_detector
{
```

br2_tf2_detector/include/br2_tf2_detector/ObstacleDetectorNode.hpp

```cpp
class ObstacleDetectorNode : public rclcpp::Node
{
public:
  ObstacleDetectorNode();

private:
  void scan_callback(sensor_msgs::msg::LaserScan::UniquePtr msg);

  rclcpp::Subscription<sensor_msgs::msg::LaserScan>::SharedPtr scan_sub_;
  std::shared_ptr<tf2_ros::StaticTransformBroadcaster> tf_broadcaster_;
};

}  // namespace br2_tf2_detector
#endif  // BR2_TF2_DETECTOR__OBSTACLEDETECTORNODE_HPP_
```

br2_tf2_detector/src/detector_main.cpp

```cpp
#include <memory>

#include "br2_tf2_detector/ObstacleDetectorNode.hpp"
#include "br2_tf2_detector/ObstacleMonitorNode.hpp"

#include "rclcpp/rclcpp.hpp"

int main(int argc, char * argv[])
{
  rclcpp::init(argc, argv);

  auto obstacle_detector = std::make_shared<br2_tf2_detector::ObstacleDetectorNode>();
  auto obstacle_monitor = std::make_shared<br2_tf2_detector::ObstacleMonitorNode>();

  rclcpp::executors::SingleThreadedExecutor executor;
  executor.add_node(obstacle_detector->get_node_base_interface());
  executor.add_node(obstacle_monitor->get_node_base_interface());

  executor.spin();

  rclcpp::shutdown();
  return 0;
}
```

br2_tf2_detector/src/detector_improved_main.cpp

```cpp
#include <memory>

#include "br2_tf2_detector/ObstacleDetectorImprovedNode.hpp"
#include "br2_tf2_detector/ObstacleMonitorNode.hpp"

#include "rclcpp/rclcpp.hpp"

int main(int argc, char * argv[])
{
  rclcpp::init(argc, argv);

  auto obstacle_detector = std::make_shared<br2_tf2_detector::
    ObstacleDetectorImprovedNode>();
  auto obstacle_monitor = std::make_shared<br2_tf2_detector::ObstacleMonitorNode>();

  rclcpp::executors::SingleThreadedExecutor executor;
  executor.add_node(obstacle_detector->get_node_base_interface());
  executor.add_node(obstacle_monitor->get_node_base_interface());

  executor.spin();

  rclcpp::shutdown();
  return 0;
}
```

br2_tf2_detector/src/br2_tf2_detector/ObstacleMonitorNode.cpp

```cpp
#include <tf2/transform_datatypes.h>
#include <tf2/LinearMath/Quaternion.h>
#include <tf2_geometry_msgs/tf2_geometry_msgs.h>

#include <memory>

#include "br2_tf2_detector/ObstacleMonitorNode.hpp"

#include "geometry_msgs/msg/transform_stamped.hpp"

#include "rclcpp/rclcpp.hpp"

namespace br2_tf2_detector
{

using namespace std::chrono_literals;

ObstacleMonitorNode::ObstacleMonitorNode()
: Node("obstacle_monitor"),
  tf_buffer_(),
  tf_listener_(tf_buffer_)
{
  marker_pub_ = create_publisher<visualization_msgs::msg::Marker>("obstacle_marker", 1);

  timer_ = create_wall_timer(
    500ms, std::bind(&ObstacleMonitorNode::control_cycle, this));
}

void
ObstacleMonitorNode::control_cycle()
{
  geometry_msgs::msg::TransformStamped robot2obstacle;

  try {
    robot2obstacle = tf_buffer_.lookupTransform(
      "base_footprint", "detected_obstacle", tf2::TimePointZero);
  } catch (tf2::TransformException & ex) {
    RCLCPP_WARN(get_logger(), "Obstacle transform not found: %s", ex.what());
    return;
  }

  double x = robot2obstacle.transform.translation.x;
  double y = robot2obstacle.transform.translation.y;
  double z = robot2obstacle.transform.translation.z;
  double theta = atan2(y, x);

  RCLCPP_INFO(
    get_logger(), "Obstacle detected at (%lf m, %lf m, , %lf m) = %lf rads",
    x, y, z, theta);

  visualization_msgs::msg::Marker obstacle_arrow;
  obstacle_arrow.header.frame_id = "base_footprint";
  obstacle_arrow.header.stamp = now();
  obstacle_arrow.type = visualization_msgs::msg::Marker::ARROW;
  obstacle_arrow.action = visualization_msgs::msg::Marker::ADD;
  obstacle_arrow.lifetime = rclcpp::Duration(1s);

  geometry_msgs::msg::Point start;
  start.x = 0.0;
  start.y = 0.0;
  start.z = 0.0;
  geometry_msgs::msg::Point end;
  end.x = x;
  end.y = y;
  end.z = z;
  obstacle_arrow.points = {start, end};

  obstacle_arrow.color.r = 1.0;
  obstacle_arrow.color.g = 0.0;
  obstacle_arrow.color.b = 0.0;
  obstacle_arrow.color.a = 1.0;
```

br2_tf2_detector/src/br2_tf2_detector/ObstacleMonitorNode.cpp

```cpp
  obstacle_arrow.scale.x = 0.02;
  obstacle_arrow.scale.y = 0.1;
  obstacle_arrow.scale.z = 0.1;

  marker_pub_->publish(obstacle_arrow);
}

}  // namespace br2_tf2_detector
```

br2_tf2_detector/src/br2_tf2_detector/ObstacleDetectorNode.cpp

```cpp
#include <memory>

#include "br2_tf2_detector/ObstacleDetectorNode.hpp"

#include "sensor_msgs/msg/laser_scan.hpp"
#include "geometry_msgs/msg/transform_stamped.hpp"

#include "rclcpp/rclcpp.hpp"

namespace br2_tf2_detector
{

using std::placeholders::_1;

ObstacleDetectorNode::ObstacleDetectorNode()
: Node("obstacle_detector")
{
  scan_sub_ = create_subscription<sensor_msgs::msg::LaserScan>(
    "input_scan", rclcpp::SensorDataQoS(),
    std::bind(&ObstacleDetectorNode::scan_callback, this, _1));

  tf_broadcaster_ = std::make_shared<tf2_ros::StaticTransformBroadcaster>(*this);
}

void
ObstacleDetectorNode::scan_callback(sensor_msgs::msg::LaserScan::UniquePtr msg)
{
  double dist = msg->ranges[msg->ranges.size() / 2];

  if (!std::isinf(dist)) {
    geometry_msgs::msg::TransformStamped detection_tf;

    detection_tf.header = msg->header;
    detection_tf.child_frame_id = "detected_obstacle";
    detection_tf.transform.translation.x = msg->ranges[msg->ranges.size() / 2];

    tf_broadcaster_->sendTransform(detection_tf);
  }
}

}  // namespace br2_tf2_detector
```

br2_tf2_detector/src/br2_tf2_detector/ObstacleDetectorImprovedNode.cpp

```cpp
#include <tf2/transform_datatypes.h>
#include <tf2/LinearMath/Quaternion.h>
#include <tf2_geometry_msgs/tf2_geometry_msgs.h>

#include <memory>

#include "br2_tf2_detector/ObstacleDetectorImprovedNode.hpp"

#include "sensor_msgs/msg/laser_scan.hpp"
#include "geometry_msgs/msg/transform_stamped.hpp"

#include "rclcpp/rclcpp.hpp"
```

br2_tf2_detector/src/br2_tf2_detector/ObstacleDetectorImprovedNode.cpp

```cpp
namespace br2_tf2_detector
{

using std::placeholders::_1;
using namespace std::chrono_literals;

ObstacleDetectorImprovedNode::ObstacleDetectorImprovedNode()
: Node("obstacle_detector_improved"),
  tf_buffer_(),
  tf_listener_(tf_buffer_)
{
  scan_sub_ = create_subscription<sensor_msgs::msg::LaserScan>(
    "input_scan", rclcpp::SensorDataQoS(),
    std::bind(&ObstacleDetectorImprovedNode::scan_callback, this, _1));

  tf_broadcaster_ = std::make_shared<tf2_ros::StaticTransformBroadcaster>(*this);
}

void
ObstacleDetectorImprovedNode::scan_callback(sensor_msgs::msg::LaserScan::UniquePtr msg)
{
  double dist = msg->ranges[msg->ranges.size() / 2];

  if (!std::isinf(dist)) {
    tf2::Transform laser2object;
    laser2object.setOrigin(tf2::Vector3(dist, 0.0, 0.0));
    laser2object.setRotation(tf2::Quaternion(0.0, 0.0, 0.0, 1.0));

    geometry_msgs::msg::TransformStamped odom2laser_msg;
    tf2::Stamped<tf2::Transform> odom2laser;
    try {
      odom2laser_msg = tf_buffer_.lookupTransform(
        "odom", "base_laser_link",
        tf2::timeFromSec(rclcpp::Time(msg->header.stamp).seconds()));
      tf2::fromMsg(odom2laser_msg, odom2laser);
    } catch (tf2::TransformException & ex) {
      RCLCPP_WARN(get_logger(), "Obstacle transform not found: %s", ex.what());
      return;
    }

    tf2::Transform odom2object = odom2laser * laser2object;

    geometry_msgs::msg::TransformStamped odom2object_msg;
    odom2object_msg.transform = tf2::toMsg(odom2object);

    odom2object_msg.header.stamp = msg->header.stamp;
    odom2object_msg.header.frame_id = "odom";
    odom2object_msg.child_frame_id = "detected_obstacle";

    tf_broadcaster_->sendTransform(odom2object_msg);
  }
}

}  // namespace br2_tf2_detector
```

A.5 BR2_VFF_AVOIDANCE 软件包

Package br2_vff_avoidance

```
br2_vff_avoidance
├── CMakeLists.txt
├── include
│   └── br2_vff_avoidance
│       └── AvoidanceNode.hpp
├── launch
```

```
Package br2_vff_avoidance
    └── avoidance_vff.launch.py
├── package.xml
├── src
│   ├── avoidance_vff_main.cpp
│   └── br2_vff_avoidance
│       └── AvoidanceNode.cpp
└── tests
    ├── CMakeLists.txt
    └── vff_test.cpp
```

br2_vff_avoidance/CMakeLists.txt

```cmake
cmake_minimum_required(VERSION 3.5)
project(br2_vff_avoidance)

set(CMAKE_CXX_STANDARD 17)
set(CMAKE_BUILD_TYPE Debug)

find_package(ament_cmake REQUIRED)
find_package(rclcpp REQUIRED)
find_package(sensor_msgs REQUIRED)
find_package(geometry_msgs REQUIRED)
find_package(visualization_msgs REQUIRED)

set(dependencies
  rclcpp
  sensor_msgs
  geometry_msgs
  visualization_msgs
)

include_directories(include)

add_library(${PROJECT_NAME} SHARED src/br2_vff_avoidance/AvoidanceNode.cpp)
ament_target_dependencies(${PROJECT_NAME} ${dependencies})

add_executable(avoidance_vff src/avoidance_vff_main.cpp)
ament_target_dependencies(avoidance_vff ${dependencies})
target_link_libraries(avoidance_vff ${PROJECT_NAME})

install(TARGETS
  ${PROJECT_NAME}
  avoidance_vff
  ARCHIVE DESTINATION lib
  LIBRARY DESTINATION lib
  RUNTIME DESTINATION lib/${PROJECT_NAME}
)

install(DIRECTORY launch DESTINATION share/${PROJECT_NAME})

if(BUILD_TESTING)
  find_package(ament_lint_auto REQUIRED)
  ament_lint_auto_find_test_dependencies()

  set(ament_cmake_cpplint_FOUND TRUE)
  ament_lint_auto_find_test_dependencies()

  find_package(ament_cmake_gtest REQUIRED)
  add_subdirectory(tests)
endif()

ament_export_dependencies(${dependencies})
ament_package()
```

br2_vff_avoidance/launch/avoidance_vff.launch.py

```python
from launch import LaunchDescription
from launch_ros.actions import Node
```

br2_vff_avoidance/launch/avoidance_vff.launch.py

```python
def generate_launch_description():

    vff_avoidance_cmd = Node(
      package='br2_vff_avoidance',
      executable='avoidance_vff',
      parameters=[{
        'use_sim_time': True
      }],
      remappings=[
        ('input_scan', '/scan_raw'),
        ('output_vel', '/nav_vel')
      ],
      output='screen'
    )

    ld = LaunchDescription()
    ld.add_action(vff_avoidance_cmd)

    return ld
```

br2_vff_avoidance/package.xml

```xml
<?xml version="1.0"?>
<?xml-model href="http://download.ros.org/schema/package_format3.xsd"
  schematypens="http://www.w3.org/2001/XMLSchema"?>
<package format="3">
  <name>br2_vff_avoidance</name>
  <version>0.1.0</version>
  <description>VFF Avoidance package</description>
  <maintainer email="fmrico@gmail.com">Francisco Martín</maintainer>
  <license>Apache 2.0</license>

  <buildtool_depend>ament_cmake</buildtool_depend>

  <depend>rclcpp</depend>
  <depend>geometry_msgs</depend>
  <depend>sensor_msgs</depend>
  <depend>visualization_msgs</depend>

  <test_depend>ament_lint_auto</test_depend>
  <test_depend>ament_lint_common</test_depend>
  <test_depend>ament_cmake_gtest</test_depend>

  <export>
    <build_type>ament_cmake</build_type>
  </export>
</package>
```

br2_vff_avoidance/include/br2_vff_avoidance/AvoidanceNode.hpp

```cpp
#ifndef BR2_VFF_AVOIDANCE__AVOIDANCENODE_HPP_
#define BR2_VFF_AVOIDANCE__AVOIDANCENODE_HPP_

#include <memory>
#include <vector>

#include "geometry_msgs/msg/twist.hpp"
#include "sensor_msgs/msg/laser_scan.hpp"
#include "visualization_msgs/msg/marker_array.hpp"

#include "rclcpp/rclcpp.hpp"

namespace br2_vff_avoidance
{

struct VFFVectors
{
  std::vector<float> attractive;
```

`br2_vff_avoidance/include/br2_vff_avoidance/AvoidanceNode.hpp`

```cpp
  std::vector<float> repulsive;
  std::vector<float> result;
};

typedef enum {RED, GREEN, BLUE, NUM_COLORS} VFFColor;

class AvoidanceNode : public rclcpp::Node
{
public:
  AvoidanceNode();

  void scan_callback(sensor_msgs::msg::LaserScan::UniquePtr msg);
  void control_cycle();

protected:
  VFFVectors get_vff(const sensor_msgs::msg::LaserScan & scan);

  visualization_msgs::msg::MarkerArray get_debug_vff(const VFFVectors & vff_vectors);
  visualization_msgs::msg::Marker make_marker(
    const std::vector<float> & vector, VFFColor vff_color);

private:
  rclcpp::Publisher<geometry_msgs::msg::Twist>::SharedPtr vel_pub_;
  rclcpp::Publisher<visualization_msgs::msg::MarkerArray>::SharedPtr vff_debug_pub_;
  rclcpp::Subscription<sensor_msgs::msg::LaserScan>::SharedPtr scan_sub_;
  rclcpp::TimerBase::SharedPtr timer_;

  sensor_msgs::msg::LaserScan::UniquePtr last_scan_;
};

}  // namespace br2_vff_avoidance

#endif  // BR2_VFF_AVOIDANCE__AVOIDANCENODE_HPP_
```

`br2_vff_avoidance/src/br2_vff_avoidance/AvoidanceNode.cpp`

```cpp
#include <memory>
#include <utility>
#include <algorithm>
#include <vector>

#include "geometry_msgs/msg/twist.hpp"
#include "sensor_msgs/msg/laser_scan.hpp"
#include "visualization_msgs/msg/marker_array.hpp"

#include "br2_vff_avoidance/AvoidanceNode.hpp"

#include "rclcpp/rclcpp.hpp"

using std::placeholders::_1;
using namespace std::chrono_literals;

namespace br2_vff_avoidance
{

AvoidanceNode::AvoidanceNode()
: Node("avoidance_vff")
{
  vel_pub_ = create_publisher<geometry_msgs::msg::Twist>("output_vel", 100);
  vff_debug_pub_ = create_publisher<visualization_msgs::msg::MarkerArray>("vff_debug",
    100);
  scan_sub_ = create_subscription<sensor_msgs::msg::LaserScan>(
    "input_scan", rclcpp::SensorDataQoS(), std::bind(&AvoidanceNode::scan_callback,
    this, _1));

  timer_ = create_wall_timer(50ms, std::bind(&AvoidanceNode::control_cycle, this));
}
```

br2_vff_avoidance/src/br2_vff_avoidance/AvoidanceNode.cpp

```cpp
void
AvoidanceNode::scan_callback(sensor_msgs::msg::LaserScan::UniquePtr msg)
{
  last_scan_ = std::move(msg);
}

void
AvoidanceNode::control_cycle()
{
  // Skip cycle if no valid recent scan available
  if (last_scan_ == nullptr || (now() - last_scan_->header.stamp) > 1s) {
    return;
  }

  // Get VFF vectors
  const VFFVectors & vff = get_vff(*last_scan_);

  // Use result vector to calculate output speed
  const auto & v = vff.result;
  double angle = atan2(v[1], v[0]);
  double module = sqrt(v[0] * v[0] + v[1] * v[1]);

  // Create ouput message, controlling speed limits
  geometry_msgs::msg::Twist vel;
  vel.linear.x = std::clamp(module, 0.0, 0.3);    // truncate linear vel to [0.0, 0.3] m/s
  vel.angular.z = std::clamp(angle, -0.5, 0.5);   // truncate rot vel to [-0.5, 0.5] rad/s

  vel_pub_->publish(vel);

  // Produce debug information, if any interested
  if (vff_debug_pub_->get_subscription_count() > 0) {
    vff_debug_pub_->publish(get_debug_vff(vff));
  }
}

VFFVectors
AvoidanceNode::get_vff(const sensor_msgs::msg::LaserScan & scan)
{
  // This is the obstacle radious in which an obstacle affects the robot
  const float OBSTACLE_DISTANCE = 1.0;

  // Init vectors
  VFFVectors vff_vector;
  vff_vector.attractive = {OBSTACLE_DISTANCE, 0.0};   // Robot wants to go forward
  vff_vector.repulsive = {0.0, 0.0};
  vff_vector.result = {0.0, 0.0};

  // Get the index of nearest obstacle
  int min_idx = std::min_element(scan.ranges.begin(), scan.ranges.end()) -
    scan.ranges.begin();

  // Get the distance to nearest obstacle
  float distance_min = scan.ranges[min_idx];

  // If the obstacle is in the area that affects the robot, calculate repulsive vector
  if (distance_min < OBSTACLE_DISTANCE) {
    float angle = scan.angle_min + scan.angle_increment * min_idx;

    float oposite_angle = angle + M_PI;
    // The module of the vector is inverse to the distance to the obstacle
    float complementary_dist = OBSTACLE_DISTANCE - distance_min;

    // Get cartesian (x, y) components from polar (angle, distance)
    vff_vector.repulsive[0] = cos(oposite_angle) * complementary_dist;
    vff_vector.repulsive[1] = sin(oposite_angle) * complementary_dist;
  }

  // Calculate resulting vector adding attractive and repulsive vectors
  vff_vector.result[0] = (vff_vector.repulsive[0] + vff_vector.attractive[0]);
  vff_vector.result[1] = (vff_vector.repulsive[1] + vff_vector.attractive[1]);

  return vff_vector;
}
```

br2_vff_avoidance/src/br2_vff_avoidance/AvoidanceNode.cpp

```cpp
visualization_msgs::msg::MarkerArray
AvoidanceNode::get_debug_vff(const VFFVectors & vff_vectors)
{
  visualization_msgs::msg::MarkerArray marker_array;

  marker_array.markers.push_back(make_marker(vff_vectors.attractive, BLUE));
  marker_array.markers.push_back(make_marker(vff_vectors.repulsive, RED));
  marker_array.markers.push_back(make_marker(vff_vectors.result, GREEN));

  return marker_array;
}

visualization_msgs::msg::Marker
AvoidanceNode::make_marker(const std::vector<float> & vector, VFFColor vff_color)
{
  visualization_msgs::msg::Marker marker;

  marker.header.frame_id = "base_footprint";
  marker.header.stamp = now();
  marker.type = visualization_msgs::msg::Marker::ARROW;
  marker.id = visualization_msgs::msg::Marker::ADD;

  geometry_msgs::msg::Point start;
  start.x = 0.0;
  start.y = 0.0;
  geometry_msgs::msg::Point end;
  start.x = vector[0];
  start.y = vector[1];
  marker.points = {end, start};

  marker.scale.x = 0.05;
  marker.scale.y = 0.1;

  switch (vff_color) {
    case RED:
      marker.id = 0;
      marker.color.r = 1.0;
      break;
    case GREEN:
      marker.id = 1;
      marker.color.g = 1.0;
      break;
    case BLUE:
      marker.id = 2;
      marker.color.b = 1.0;
      break;
  }
  marker.color.a = 1.0;

  return marker;
}

}  // namespace br2_vff_avoidance
```

br2_vff_avoidance/src/avoidance_vff_main.cpp

```cpp
#include <memory>

#include "br2_vff_avoidance/AvoidanceNode.hpp"
#include "rclcpp/rclcpp.hpp"

int main(int argc, char * argv[])
{
  rclcpp::init(argc, argv);

  auto avoidance_node = std::make_shared<br2_vff_avoidance::AvoidanceNode>();
  rclcpp::spin(avoidance_node);

  rclcpp::shutdown();

  return 0;
}
```

br2_vff_avoidance/tests/vff_test.cpp

```cpp
#include <limits>
#include <vector>
#include <memory>

#include "sensor_msgs/msg/laser_scan.hpp"
#include "br2_vff_avoidance/AvoidanceNode.hpp"

#include "gtest/gtest.h"

using namespace std::chrono_literals;

class AvoidanceNodeTest : public br2_vff_avoidance::AvoidanceNode
{
public:
  br2_vff_avoidance::VFFVectors
  get_vff_test(const sensor_msgs::msg::LaserScan & scan)
  {
    return get_vff(scan);
  }

  visualization_msgs::msg::MarkerArray
  get_debug_vff_test(const br2_vff_avoidance::VFFVectors & vff_vectors)
  {
    return get_debug_vff(vff_vectors);
  }
};

sensor_msgs::msg::LaserScan get_scan_test_1(rclcpp::Time ts)
{
  sensor_msgs::msg::LaserScan ret;
  ret.header.stamp = ts;
  ret.angle_min = -M_PI;
  ret.angle_max = M_PI;
  ret.angle_increment = 2.0 * M_PI / 16.0;
  ret.ranges = std::vector<float>(16, std::numeric_limits<float>::infinity());

  return ret;
}

sensor_msgs::msg::LaserScan get_scan_test_2(rclcpp::Time ts)
{
  sensor_msgs::msg::LaserScan ret;
  ret.header.stamp = ts;
  ret.angle_min = -M_PI;
  ret.angle_max = M_PI;
  ret.angle_increment = 2.0 * M_PI / 16.0;
  ret.ranges = std::vector<float>(16, 0.0);

  return ret;
}

sensor_msgs::msg::LaserScan get_scan_test_3(rclcpp::Time ts)
{
  sensor_msgs::msg::LaserScan ret;
  ret.header.stamp = ts;
  ret.angle_min = -M_PI;
  ret.angle_max = M_PI;
  ret.angle_increment = 2.0 * M_PI / 16.0;
  ret.ranges = std::vector<float>(16, 5.0);
  ret.ranges[2] = 0.3;

  return ret;
}

sensor_msgs::msg::LaserScan get_scan_test_4(rclcpp::Time ts)
{
  sensor_msgs::msg::LaserScan ret;
  ret.header.stamp = ts;
  ret.angle_min = -M_PI;
  ret.angle_max = M_PI;
  ret.angle_increment = 2.0 * M_PI / 16.0;
```

```
br2_vff_avoidance/tests/vff_test.cpp
    ret.ranges = std::vector<float>(16, 5.0);
    ret.ranges[6] = 0.3;

    return ret;
}
sensor_msgs::msg::LaserScan get_scan_test_5(rclcpp::Time ts)
{
  sensor_msgs::msg::LaserScan ret;
  ret.header.stamp = ts;
  ret.angle_min = -M_PI;
  ret.angle_max = M_PI;
ret.angle_increment = 2.0 * M_PI / 16.0;
  ret.ranges = std::vector<float>(16, 5.0);
  ret.ranges[10] = 0.3;

  return ret;
}

sensor_msgs::msg::LaserScan get_scan_test_6(rclcpp::Time ts)
{
  sensor_msgs::msg::LaserScan ret;
  ret.header.stamp = ts;
  ret.angle_min = -M_PI;
  ret.angle_max = M_PI;
  ret.angle_increment = 2.0 * M_PI / 16.0;
  ret.ranges = std::vector<float>(16, 0.5);
  ret.ranges[10] = 0.3;

  return ret;
}

sensor_msgs::msg::LaserScan get_scan_test_7(rclcpp::Time ts)
{
  sensor_msgs::msg::LaserScan ret;
  ret.header.stamp = ts;
  ret.angle_min = -M_PI;
  ret.angle_max = M_PI;
  ret.angle_increment = 2.0 * M_PI / 16.0;
  ret.ranges = std::vector<float>(16, 5.0);
  ret.ranges[14] = 0.3;

  return ret;
}

sensor_msgs::msg::LaserScan get_scan_test_8(rclcpp::Time ts)
{
  sensor_msgs::msg::LaserScan ret;
  ret.header.stamp = ts;
  ret.angle_min = -M_PI;
  ret.angle_max = M_PI;
  ret.angle_increment = 2.0 * M_PI / 16.0;
  ret.ranges = std::vector<float>(16, 5.0);
  ret.ranges[8] = 0.01;

  return ret;
}

TEST(vff_tests, get_vff)
{
  auto node_avoidance = AvoidanceNodeTest();

  rclcpp::Time ts = node_avoidance.now();

  auto res1 = node_avoidance.get_vff_test(get_scan_test_1(ts));
  ASSERT_EQ(res1.attractive, std::vector<float>({1.0f, 0.0f}));
  ASSERT_EQ(res1.repulsive, std::vector<float>({0.0f, 0.0f}));
  ASSERT_EQ(res1.result, std::vector<float>({1.0f, 0.0f}));

  auto res2 = node_avoidance.get_vff_test(get_scan_test_2(ts));
```

br2_vff_avoidance/tests/vff_test.cpp

```cpp
    ASSERT_EQ(res2.attractive, std::vector<float>({1.0f, 0.0f}));
    ASSERT_NEAR(res2.repulsive[0], 1.0f, 0.00001f);
    ASSERT_NEAR(res2.repulsive[1], 0.0f, 0.00001f);
    ASSERT_NEAR(res2.result[0], 2.0f, 0.00001f);
    ASSERT_NEAR(res2.result[1], 0.0f, 0.00001f);

    auto res3 = node_avoidance.get_vff_test(get_scan_test_3(ts));
    ASSERT_EQ(res3.attractive, std::vector<float>({1.0f, 0.0f}));
    ASSERT_GT(res3.repulsive[0], 0.0f);
    ASSERT_GT(res3.repulsive[1], 0.0f);
    ASSERT_GT(atan2(res3.repulsive[1], res3.repulsive[0]), 0.1);
    ASSERT_LT(atan2(res3.repulsive[1], res3.repulsive[0]), M_PI_2);
    ASSERT_GT(atan2(res3.result[1], res3.result[0]), 0.1);
    ASSERT_LT(atan2(res3.result[1], res3.result[0]), M_PI_2);

    auto res4 = node_avoidance.get_vff_test(get_scan_test_4(ts));
    ASSERT_EQ(res4.attractive, std::vector<float>({1.0f, 0.0f}));
    ASSERT_LT(res4.repulsive[0], 0.0f);
    ASSERT_GT(res4.repulsive[1], 0.0f);
    ASSERT_GT(atan2(res4.repulsive[1], res4.repulsive[0]), M_PI_2);
    ASSERT_LT(atan2(res4.repulsive[1], res4.repulsive[0]), M_PI);
    ASSERT_GT(atan2(res4.result[1], res4.result[0]), 0.0);
    ASSERT_LT(atan2(res4.result[1], res4.result[0]), M_PI_2);

    auto res5 = node_avoidance.get_vff_test(get_scan_test_5(ts));
    ASSERT_EQ(res5.attractive, std::vector<float>({1.0f, 0.0f}));
    ASSERT_LT(res5.repulsive[0], 0.0f);
    ASSERT_LT(res5.repulsive[1], 0.0f);
    ASSERT_GT(atan2(res5.repulsive[1], res5.repulsive[0]), -M_PI);
    ASSERT_LT(atan2(res5.repulsive[1], res5.repulsive[0]), -M_PI_2);
    ASSERT_LT(atan2(res5.result[1], res5.result[0]), 0.0);
    ASSERT_GT(atan2(res5.result[1], res5.result[0]), -M_PI_2);

    auto res6 = node_avoidance.get_vff_test(get_scan_test_6(ts));
    ASSERT_EQ(res6.attractive, std::vector<float>({1.0f, 0.0f}));
    ASSERT_GT(res6.repulsive[0], 0.0f);
    ASSERT_LT(res6.repulsive[1], 0.0f);
    ASSERT_GT(atan2(res6.repulsive[1], res6.repulsive[0]), -M_PI);
    ASSERT_LT(atan2(res6.repulsive[1], res6.repulsive[0]), -M_PI_2);
    ASSERT_LT(atan2(res6.result[1], res6.result[0]), 0.0);
    ASSERT_GT(atan2(res6.result[1], res6.result[0]), -M_PI_2);

    auto res7 = node_avoidance.get_vff_test(get_scan_test_7(ts));
    ASSERT_EQ(res7.attractive, std::vector<float>({1.0f, 0.0f}));
    ASSERT_GT(res7.repulsive[0], 0.0f);
    ASSERT_LT(res7.repulsive[1], 0.0f);
    ASSERT_LT(atan2(res7.repulsive[1], res7.repulsive[0]), 0.0f);
    ASSERT_GT(atan2(res7.repulsive[1], res7.repulsive[0]), -M_PI_2);
    ASSERT_LT(atan2(res7.result[1], res7.result[0]), 0.0);
    ASSERT_GT(atan2(res7.result[1], res7.result[0]), -M_PI_2);

    auto res8 = node_avoidance.get_vff_test(get_scan_test_8(ts));
    ASSERT_EQ(res8.attractive, std::vector<float>({1.0f, 0.0f}));
    ASSERT_NEAR(res8.repulsive[0], -1.0f, 0.1f);
    ASSERT_NEAR(res8.repulsive[1], 0.0f, 0.0001f);
    ASSERT_NEAR(res8.result[0], 0.0f, 0.01f);
    ASSERT_NEAR(res8.result[1], 0.0f, 0.01f);
}

TEST(vff_tests, ouput_vels)
{
    auto node_avoidance = std::make_shared<AvoidanceNodeTest>();

    // Create a testing node with a scan publisher and a speed subscriber
    auto test_node = rclcpp::Node::make_shared("test_node");
    auto scan_pub = test_node->create_publisher<sensor_msgs::msg::LaserScan>(
      "input_scan", 100);

    geometry_msgs::msg::Twist last_vel;
    auto vel_sub = test_node->create_subscription<geometry_msgs::msg::Twist>(
      "output_vel", 1, [&last_vel](geometry_msgs::msg::Twist::SharedPtr msg) {
```

```
br2_vff_avoidance/tests/vff_test.cpp
      last_vel = *msg;
    });

  ASSERT_EQ(vel_sub->get_publisher_count(), 1);
  ASSERT_EQ(scan_pub->get_subscription_count(), 1);

  rclcpp::Rate rate(30);
  rclcpp::executors::SingleThreadedExecutor executor;
  executor.add_node(node_avoidance);
  executor.add_node(test_node);

  // Test for scan test #1
  auto start = node_avoidance->now();
  while (rclcpp::ok() && (node_avoidance->now() - start) < 1s) {
    scan_pub->publish(get_scan_test_1(node_avoidance->now()));
    executor.spin_some();
    rate.sleep();
  }
  ASSERT_NEAR(last_vel.linear.x, 0.3f, 0.0001f);
  ASSERT_NEAR(last_vel.angular.z, 0.0f, 0.0001f);

  // Test for scan test #2
  start = node_avoidance->now();
  while (rclcpp::ok() && (node_avoidance->now() - start) < 1s) {
    scan_pub->publish(get_scan_test_2(node_avoidance->now()));
    executor.spin_some();
    rate.sleep();
  }
  ASSERT_NEAR(last_vel.linear.x, 0.3f, 0.0001f);
  ASSERT_NEAR(last_vel.angular.z, 0.0f, 0.0001f);

  // Test for scan test #3
  start = node_avoidance->now();
  while (rclcpp::ok() && (node_avoidance->now() - start) < 1s) {
    scan_pub->publish(get_scan_test_3(node_avoidance->now()));
    executor.spin_some();
    rate.sleep();
  }
  ASSERT_LT(last_vel.linear.x, 0.3f);
  ASSERT_GT(last_vel.linear.x, 0.0f);
  ASSERT_GT(last_vel.angular.z, 0.0f);
  ASSERT_LT(last_vel.angular.z, M_PI_2);

  // Test for scan test #4
  start = node_avoidance->now();
  while (rclcpp::ok() && (node_avoidance->now() - start) < 1s) {
    scan_pub->publish(get_scan_test_4(node_avoidance->now()));
    executor.spin_some();
    rate.sleep();
  }
  ASSERT_LT(last_vel.linear.x, 0.3f);
  ASSERT_GT(last_vel.linear.x, 0.0f);
  ASSERT_GT(last_vel.angular.z, 0.0f);
  ASSERT_LT(last_vel.angular.z, M_PI_2);

  // Test for scan test #5
  start = node_avoidance->now();
  while (rclcpp::ok() && (node_avoidance->now() - start) < 1s) {
    scan_pub->publish(get_scan_test_5(node_avoidance->now()));
    executor.spin_some();
    rate.sleep();
  }
  ASSERT_LT(last_vel.linear.x, 0.3f);
  ASSERT_GT(last_vel.linear.x, 0.0f);
  ASSERT_LT(last_vel.angular.z, 0.0f);
  ASSERT_GT(last_vel.angular.z, -M_PI_2);

  // Test for scan test #6
  start = node_avoidance->now();
  while (rclcpp::ok() && (node_avoidance->now() - start) < 1s) {
    scan_pub->publish(get_scan_test_6(node_avoidance->now()));
```

br2_vff_avoidance/tests/vff_test.cpp

```cpp
    executor.spin_some();
    rate.sleep();
  }
  ASSERT_LT(last_vel.linear.x, 0.3f);
  ASSERT_GT(last_vel.linear.x, 0.0f);
  ASSERT_LT(last_vel.angular.z, 0.0f);
  ASSERT_GT(last_vel.angular.z, -M_PI_2);

  // Test for scan test #7
  start = node_avoidance->now();
  while (rclcpp::ok() && (node_avoidance->now() - start) < 1s) {
    scan_pub->publish(get_scan_test_7(node_avoidance->now()));
    executor.spin_some();
    rate.sleep();
  }
  ASSERT_LT(last_vel.linear.x, 0.3f);
  ASSERT_GT(last_vel.linear.x, 0.0f);
  ASSERT_LT(last_vel.angular.z, 0.0f);
  ASSERT_GT(last_vel.angular.z, -M_PI_2);

  // Test for scan test #8
  start = node_avoidance->now();
  while (rclcpp::ok() && (node_avoidance->now() - start) < 2s) {
    scan_pub->publish(get_scan_test_8(node_avoidance->now()));
    executor.spin_some();
    rate.sleep();
  }
  ASSERT_NEAR(last_vel.linear.x, 0.0f, 0.1f);
  ASSERT_LT(last_vel.angular.z, 0.0f);
  ASSERT_GT(last_vel.angular.z, -M_PI_2);

  // Test for stooping when scan is too old
  last_vel = geometry_msgs::msg::Twist();
  while (rclcpp::ok() && (node_avoidance->now() - start) < 3s) {
    scan_pub->publish(get_scan_test_6(start));
    executor.spin_some();
    rate.sleep();
  }
  ASSERT_NEAR(last_vel.linear.x, 0.0f, 0.01f);
  ASSERT_NEAR(last_vel.angular.z, 0.0f, 0.01f);
}

int main(int argc, char ** argv)
{
  rclcpp::init(argc, argv);

  testing::InitGoogleTest(&argc, argv);
  return RUN_ALL_TESTS();
}
```

br2_vff_avoidance/tests/CMakeLists.txt

```
ament_add_gtest(vff_test vff_test.cpp)
ament_target_dependencies(vff_test ${dependencies})
target_link_libraries(vff_test ${PROJECT_NAME})
```

A.6　BR2_TRACKING_MSGS 软件包

Package br2_tracking_msgs

```
br2_tracking_msgs
├── CMakeLists.txt
├── msg
│   └── PanTiltCommand.msg
```

```
br2_tracking_msgs/CMakeLists.txt
```

```cmake
project(br2_tracking_msgs)

cmake_minimum_required(VERSION 3.5)

find_package(ament_cmake REQUIRED)
find_package(builtin_interfaces REQUIRED)
find_package(rosidl_default_generators REQUIRED)

rosidl_generate_interfaces(${PROJECT_NAME}
  "msg/PanTiltCommand.msg"
  DEPENDENCIES builtin_interfaces
)

ament_export_dependencies(rosidl_default_runtime)
ament_package()
```

```
br2_tracking_msgs/package.xml
```

```xml
<?xml version="1.0"?>
<?xml-model href="http://download.ros.org/schema/package_format3.xsd"
  schematypens="http://www.w3.org/2001/XMLSchema"?>
<package format="3">
  <name>br2_tracking_msgs</name>
  <version>0.0.0</version>

  <description>Messages for br2_tracking</description>

  <maintainer email="fmrico@gmail.com">Francisco Martin Rico</maintainer>

  <license>Apache License, Version 2.0</license>

  <buildtool_depend>ament_cmake</buildtool_depend>

  <depend>rclcpp</depend>
  <depend>builtin_interfaces</depend>
  <depend>rosidl_default_generators</depend>

  <member_of_group>rosidl_interface_packages</member_of_group>

  <export>
    <build_type>ament_cmake</build_type>
  </export>
</package>
```

A.7　BR2_TRACKING 软件包

```
Package br2_tracking
br2_tracking
├── CMakeLists.txt
├── config
│   └── detector.yaml
├── include
│   └── br2_tracking
│       ├── HeadController.hpp
│       ├── ObjectDetector.hpp
│       └── PIDController.hpp
├── launch
│   └── tracking.launch.py
├── package.xml
├── src
│   └── br2_tracking
│       ├── HeadController.cpp
│       ├── ObjectDetector.cpp
```

Package br2_tracking

```
        └── PIDController.cpp
       ├── object_detector_main.cpp
       └── object_tracker_main.cpp
  ── tests
     ├── CMakeLists.txt
```

br2_tracking/CMakeLists.txt

```
cmake_minimum_required(VERSION 3.5)
project(br2_tracking)

set(CMAKE_CXX_STANDARD 17)
set(CMAKE_BUILD_TYPE Debug)

find_package(ament_cmake REQUIRED)
find_package(rclcpp REQUIRED)
find_package(rclcpp_lifecycle REQUIRED)
find_package(br2_tracking_msgs REQUIRED)
find_package(sensor_msgs REQUIRED)
find_package(geometry_msgs REQUIRED)
find_package(vision_msgs REQUIRED)
find_package(control_msgs REQUIRED)
find_package(image_transport REQUIRED)
find_package(cv_bridge REQUIRED)

find_package(OpenCV REQUIRED)

set(dependencies
  rclcpp
  rclcpp_lifecycle
  br2_tracking_msgs
  sensor_msgs
  geometry_msgs
  vision_msgs
  control_msgs
  image_transport
  cv_bridge
  OpenCV
)

include_directories(include)

add_library(${PROJECT_NAME} SHARED
  src/br2_tracking/ObjectDetector.cpp
  src/br2_tracking/HeadController.cpp
  src/br2_tracking/PIDController.cpp
)
ament_target_dependencies(${PROJECT_NAME} ${dependencies})

add_executable(object_detector src/object_detector_main.cpp)
ament_target_dependencies(object_detector ${dependencies})
target_link_libraries(object_detector ${PROJECT_NAME})

add_executable(object_tracker src/object_tracker_main.cpp)
ament_target_dependencies(object_tracker ${dependencies})
target_link_libraries(object_tracker ${PROJECT_NAME})

install(TARGETS
  ${PROJECT_NAME}
  object_detector
  object_tracker
  ARCHIVE DESTINATION lib
  LIBRARY DESTINATION lib
  RUNTIME DESTINATION lib/${PROJECT_NAME}
)

install(
  DIRECTORY include
```

```
br2_tracking/CMakeLists.txt
```
```
    DESTINATION include
)
install(DIRECTORY launch config DESTINATION share/${PROJECT_NAME})

if(BUILD_TESTING)
  find_package(ament_lint_auto REQUIRED)
  ament_lint_auto_find_test_dependencies()

  set(ament_cmake_cpplint_FOUND TRUE)
  ament_lint_auto_find_test_dependencies()

  find_package(ament_cmake_gtest REQUIRED)
  add_subdirectory(tests)
endif()

ament_export_include_directories(include)
ament_export_libraries(${PROJECT_NAME})
ament_export_dependencies(${dependencies})
ament_package()
```

```
br2_tracking/launch/tracking.launch.py
```
```python
import os

from ament_index_python.packages import get_package_share_directory

from launch import LaunchDescription
from launch_ros.actions import Node


def generate_launch_description():

    params_file = os.path.join(
        get_package_share_directory('br2_tracking'),
        'config',
        'detector.yaml'
        )

    object_tracker_cmd = Node(
        package='br2_tracking',
        executable='object_tracker',
        parameters=[{
          'use_sim_time': True
        }, params_file],
        remappings=[
          ('input_image', '/head_front_camera/rgb/image_raw'),
          ('joint_state', '/head_controller/state'),
          ('joint_command', '/head_controller/joint_trajectory')
        ],
        output='screen'
    )

    ld = LaunchDescription()

    # Add any actions
    ld.add_action(object_tracker_cmd)

    return ld
```

```
br2_tracking/package.xml
```
```xml
<?xml version="1.0"?>
<?xml-model href="http://download.ros.org/schema/package_format3.xsd"
  schematypens="http://www.w3.org/2001/XMLSchema"?>
<package format="3">
  <name>br2_tracking</name>
```

br2_tracking/package.xml

```xml
<version>0.1.0</version>
<description>Mastering Hierarchical Finite State Machines in ROS 2 </description>
<maintainer email="fmrico@gmail.com">Francisco Martín</maintainer>
<license>Apache 2.0</license>

<buildtool_depend>ament_cmake</buildtool_depend>

<depend>rclcpp</depend>
<depend>rclcpp_lifecycle</depend>
<depend>geometry_msgs</depend>
<depend>br2_tracking_msgs</depend>
<depend>sensor_msgs</depend>
<depend>vision_msgs</depend>
<depend>control_msgs</depend>
<depend>image_transport</depend>
<depend>cv_bridge</depend>

<test_depend>ament_lint_auto</test_depend>
<test_depend>ament_lint_common</test_depend>
<test_depend>ament_cmake_gtest</test_depend>

<export>
  <build_type>ament_cmake</build_type>
</export>
</package>
```

br2_tracking/include/br2_tracking/PIDController.hpp

```cpp
#ifndef BR2_TRACKING__PIDCONTROLLER_HPP_
#define BR2_TRACKING__PIDCONTROLLER_HPP_

#include <cmath>

namespace br2_tracking
{

class PIDController
{
public:
  PIDController(double min_ref, double max_ref, double min_output, double max_output);

  void set_pid(double n_KP, double n_KI, double n_KD);
  double get_output(double new_reference);

private:
  double KP_, KI_, KD_;

  double min_ref_, max_ref_;
  double min_output_, max_output_;
  double prev_error_, int_error_;
};

}  // namespace br2_tracking

#endif  // BR2_TRACKING__PIDCONTROLLER_HPP_
```

br2_tracking/include/br2_tracking/ObjectDetector.hpp

```cpp
#ifndef BR2_TRACKING__OBJECTDETECTOR_HPP_
#define BR2_TRACKING__OBJECTDETECTOR_HPP_

#include <memory>
#include <vector>

#include "vision_msgs/msg/detection2_d.hpp"
```

br2_tracking/include/br2_tracking/ObjectDetector.hpp

```cpp
#include "image_transport/image_transport.hpp"
#include "rclcpp/rclcpp.hpp"

namespace br2_tracking
{

class ObjectDetector : public rclcpp::Node
{
public:
  ObjectDetector();

  void image_callback(const sensor_msgs::msg::Image::ConstSharedPtr & msg);

private:
  image_transport::Subscriber image_sub_;
  rclcpp::Publisher<vision_msgs::msg::Detection2D>::SharedPtr detection_pub_;

  // HSV ranges for detection [h - H] [s - S] [v - V]
  std::vector<double> hsv_filter_ranges_ {0, 180, 0, 255, 0, 255};
  bool debug_ {true};
};

}  // namespace br2_tracking

#endif  // BR2_TRACKING__OBJECTDETECTOR_HPP_
```

br2_tracking/include/br2_tracking/HeadController.hpp

```cpp
#ifndef BR2_TRACKING__HEADCONTROLLER_HPP_
#define BR2_TRACKING__HEADCONTROLLER_HPP_

#include <memory>

#include "br2_tracking_msgs/msg/pan_tilt_command.hpp"
#include "control_msgs/msg/joint_trajectory_controller_state.hpp"
#include "trajectory_msgs/msg/joint_trajectory.hpp"

#include "br2_tracking/PIDController.hpp"

#include "image_transport/image_transport.hpp"

#include "rclcpp_lifecycle/lifecycle_node.hpp"
#include "rclcpp/rclcpp.hpp"

namespace br2_tracking
{

using CallbackReturn =
  rclcpp_lifecycle::node_interfaces::LifecycleNodeInterface::CallbackReturn;

class HeadController : public rclcpp_lifecycle::LifecycleNode
{
public:
  HeadController();

  CallbackReturn on_configure(const rclcpp_lifecycle::State & previous_state);
  CallbackReturn on_activate(const rclcpp_lifecycle::State & previous_state);
  CallbackReturn on_deactivate(const rclcpp_lifecycle::State & previous_state);

  void control_sycle();

  void joint_state_callback(
    control_msgs::msg::JointTrajectoryControllerState::UniquePtr msg);
  void command_callback(br2_tracking_msgs::msg::PanTiltCommand::UniquePtr msg);

private:
  rclcpp::Subscription<br2_tracking_msgs::msg::PanTiltCommand>::SharedPtr command_sub_;
  rclcpp::Subscription<control_msgs::msg::JointTrajectoryControllerState>::SharedPtr
```

br2_tracking/include/br2_tracking/HeadController.hpp

```cpp
    joint_sub_;
  rclcpp_lifecycle::LifecyclePublisher<trajectory_msgs::msg::JointTrajectory>::SharedPtr
    joint_pub_;
  rclcpp::TimerBase::SharedPtr timer_;

  control_msgs::msg::JointTrajectoryControllerState::UniquePtr last_state_;
  br2_tracking_msgs::msg::PanTiltCommand::UniquePtr last_command_;
  rclcpp::Time last_command_ts_;

  PIDController pan_pid_, tilt_pid_;
};

}  // namespace br2_tracking

#endif  // BR2_TRACKING__HEADCONTROLLER_HPP_
```

br2_tracking/config/detector.yaml

```yaml
/object_detector:
  ros__parameters:
    debug: true
    hsv_ranges:
      - 15.0
      - 20.0
      - 50.0
      - 200.0
      - 20.0
      - 200.0
```

br2_tracking/src/br2_tracking/PIDController.cpp

```cpp
#include <algorithm>

#include "br2_tracking/PIDController.hpp"

namespace br2_tracking
{

PIDController::PIDController(
  double min_ref, double max_ref, double min_output, double max_output)
{
  min_ref_ = min_ref;
  max_ref_ = max_ref;
  min_output_ = min_output;
  max_output_ = max_output;
  prev_error_ = int_error_ = 0.0;

  KP_ = 0.41;
  KI_ = 0.06;
  KD_ = 0.53;
}

void
PIDController::set_pid(double n_KP, double n_KI, double n_KD)
{
  KP_ = n_KP;
  KI_ = n_KI;
  KD_ = n_KD;
}

double
PIDController::get_output(double new_reference)
{
  double ref = new_reference;
  double output = 0.0;

  // Proportional Error
  double direction = 0.0;
```

`br2_tracking/src/br2_tracking/PIDController.cpp`

```cpp
  if (ref != 0.0) {
    direction = ref / fabs(ref);
  }

  if (fabs(ref) < min_ref_) {
    output = 0.0;
  } else if (fabs(ref) > max_ref_) {
    output = direction * max_output_;
  } else {
    output = direction * min_output_ + ref * (max_output_ - min_output_);
  }

  // Integral Error
  int_error_ = (int_error_ + output) * 2.0 / 3.0;

  // Derivative Error
  double deriv_error = output - prev_error_;
  prev_error_ = output;

  output = KP_ * output + KI_ * int_error_ + KD_ * deriv_error;

  return std::clamp(output, -max_output_, max_output_);
}

}  // namespace br2_tracking
```

`br2_tracking/src/br2_tracking/HeadController.cpp`

```cpp
#include <algorithm>
#include <utility>

#include "br2_tracking/HeadController.hpp"
#include "br2_tracking/PIDController.hpp"

#include "br2_tracking_msgs/msg/pan_tilt_command.hpp"
#include "control_msgs/msg/joint_trajectory_controller_state.hpp"
#include "trajectory_msgs/msg/joint_trajectory.hpp"

#include "rclcpp_lifecycle/lifecycle_node.hpp"
#include "rclcpp/rclcpp.hpp"

namespace br2_tracking
{

using std::placeholders::_1;
using namespace std::chrono_literals;
using CallbackReturn = rclcpp_lifecycle::node_interfaces::LifecycleNodeInterface::
  CallbackReturn;

HeadController::HeadController()
: LifecycleNode("head_tracker"),
  pan_pid_(0.0, 1.0, 0.0, 0.3),
  tilt_pid_(0.0, 1.0, 0.0, 0.3)
{
  command_sub_ = create_subscription<br2_tracking_msgs::msg::PanTiltCommand>(
    "command", 100,
    std::bind(&HeadController::command_callback, this, _1));
  joint_sub_ = create_subscription<control_msgs::msg::JointTrajectoryControllerState>(
    "joint_state", rclcpp::SensorDataQoS(),
    std::bind(&HeadController::joint_state_callback, this, _1));
  joint_pub_ = create_publisher<trajectory_msgs::msg::JointTrajectory>("joint_command",
    100);
}

CallbackReturn
HeadController::on_configure(const rclcpp_lifecycle::State & previous_state)
{
  RCLCPP_INFO(get_logger(), "HeadController configured");
```

```cpp
// br2_tracking/src/br2_tracking/HeadController.cpp
  pan_pid_.set_pid(0.4, 0.05, 0.55);
  tilt_pid_.set_pid(0.4, 0.05, 0.55);

  return CallbackReturn::SUCCESS;
}

CallbackReturn
HeadController::on_activate(const rclcpp_lifecycle::State & previous_state)
{
  RCLCPP_INFO(get_logger(), "HeadController activated");

  joint_pub_->on_activate();
  timer_ = create_wall_timer(100ms, std::bind(&HeadController::control_sycle, this));

  return CallbackReturn::SUCCESS;
}

CallbackReturn
HeadController::on_deactivate(const rclcpp_lifecycle::State & previous_state)
{
  RCLCPP_INFO(get_logger(), "HeadController deactivated");

  trajectory_msgs::msg::JointTrajectory command_msg;
  command_msg.header.stamp = now();
  command_msg.joint_names = last_state_->joint_names;
  command_msg.points.resize(1);
  command_msg.points[0].positions.resize(2);
  command_msg.points[0].velocities.resize(2);
  command_msg.points[0].accelerations.resize(2);
  command_msg.points[0].positions[0] = 0.0;
  command_msg.points[0].positions[1] = 0.0;
  command_msg.points[0].velocities[0] = 0.1;
  command_msg.points[0].velocities[1] = 0.1;
  command_msg.points[0].accelerations[0] = 0.1;
  command_msg.points[0].accelerations[1] = 0.1;
  command_msg.points[0].time_from_start = rclcpp::Duration(1s);

  joint_pub_->publish(command_msg);

  joint_pub_->on_deactivate();
  timer_ = nullptr;

  return CallbackReturn::SUCCESS;
}

void
HeadController::joint_state_callback(
  control_msgs::msg::JointTrajectoryControllerState::UniquePtr msg)
{
  last_state_ = std::move(msg);
}

void
HeadController::command_callback(br2_tracking_msgs::msg::PanTiltCommand::UniquePtr msg)
{
  last_command_ = std::move(msg);
  last_command_ts_ = now();
}

void
HeadController::control_sycle()
{
  if (last_state_ == nullptr) {return;}

  trajectory_msgs::msg::JointTrajectory command_msg;
  command_msg.header.stamp = now();
  command_msg.joint_names = last_state_->joint_names;
  command_msg.points.resize(1);
  command_msg.points[0].positions.resize(2);
  command_msg.points[0].velocities.resize(2);
  command_msg.points[0].accelerations.resize(2);
```

```
br2_tracking/src/br2_tracking/HeadController.cpp
```
```cpp
    command_msg.points[0].time_from_start = rclcpp::Duration(200ms);

    if (last_command_ == nullptr || (now() - last_command_ts_) > 100ms) {
      command_msg.points[0].positions[0] = 0.0;
      command_msg.points[0].positions[1] = 0.0;
      command_msg.points[0].velocities[0] = 0.1;
      command_msg.points[0].velocities[1] = 0.1;
      command_msg.points[0].accelerations[0] = 0.1;
      command_msg.points[0].accelerations[1] = 0.1;
      command_msg.points[0].time_from_start = rclcpp::Duration(1s);
    } else {
      double control_pan = pan_pid_.get_output(last_command_->pan);
      double control_tilt = tilt_pid_.get_output(last_command_->tilt);

      command_msg.points[0].positions[0] = last_state_->actual.positions[0] - control_pan;
      command_msg.points[0].positions[1] = last_state_->actual.positions[1] - control_tilt;

      command_msg.points[0].velocities[0] = 0.5;
      command_msg.points[0].velocities[1] = 0.5;
      command_msg.points[0].accelerations[0] = 0.5;
      command_msg.points[0].accelerations[1] = 0.5;
    }

    joint_pub_->publish(command_msg);
}

}  // namespace br2_tracking
```

```
br2_tracking/src/br2_tracking/ObjectDetector.cpp
```
```cpp
#include <vector>

#include "opencv2/opencv.hpp"
#include "cv_bridge/cv_bridge.h"

#include "br2_tracking/ObjectDetector.hpp"
#include "geometry_msgs/msg/pose2_d.hpp"

#include "image_transport/image_transport.hpp"
#include "rclcpp/rclcpp.hpp"

namespace br2_tracking
{
using std::placeholders::_1;

ObjectDetector::ObjectDetector()
: Node("object_detector")
{
  image_sub_ = image_transport::create_subscription(
    this, "input_image", std::bind(&ObjectDetector::image_callback, this, _1),
    "raw", rclcpp::SensorDataQoS().get_rmw_qos_profile());

  detection_pub_ = create_publisher<vision_msgs::msg::Detection2D>("detection", 100);

  declare_parameter("hsv_ranges", hsv_filter_ranges_);
  declare_parameter("debug", debug_);
  get_parameter("hsv_ranges", hsv_filter_ranges_);
  get_parameter("debug", debug_);
}

void
ObjectDetector::image_callback(const sensor_msgs::msg::Image::ConstSharedPtr & msg)
{
  if (detection_pub_->get_subscription_count() == 0) {return;}

  const float & h = hsv_filter_ranges_[0];
  const float & H = hsv_filter_ranges_[1];
  const float & s = hsv_filter_ranges_[2];
  const float & S = hsv_filter_ranges_[3];
```

br2_tracking/src/br2_tracking/ObjectDetector.cpp

```cpp
    const float & v = hsv_filter_ranges_[4];
    const float & V = hsv_filter_ranges_[5];

    cv_bridge::CvImagePtr cv_ptr;
    try {
      cv_ptr = cv_bridge::toCvCopy(msg, sensor_msgs::image_encodings::BGR8);
    } catch (cv_bridge::Exception & e) {
      RCLCPP_ERROR(get_logger(), "cv_bridge exception: %s", e.what());
      return;
    }

    cv::Mat img_hsv;
    cv::cvtColor(cv_ptr->image, img_hsv, cv::COLOR_BGR2HSV);

    cv::Mat1b filtered;
    cv::inRange(img_hsv, cv::Scalar(h, s, v), cv::Scalar(H, S, V), filtered);

    auto moment = cv::moments(filtered, true);
    cv::Rect bbx = cv::boundingRect(filtered);

    auto m = cv::moments(filtered, true);
    if (m.m00 < 0.000001) {return;}
    int cx = m.m10 / m.m00;
    int cy = m.m01 / m.m00;

    vision_msgs::msg::Detection2D detection_msg;
    detection_msg.header = msg->header;
    detection_msg.bbox.size_x = bbx.width;
    detection_msg.bbox.size_y = bbx.height;
    detection_msg.bbox.center.x = cx;
    detection_msg.bbox.center.y = cy;
    detection_msg.source_img = *cv_ptr->toImageMsg();
    detection_pub_->publish(detection_msg);

    if (debug_) {
      cv::rectangle(cv_ptr->image, bbx, cv::Scalar(0, 0, 255), 3);
      cv::circle(cv_ptr->image, cv::Point(cx, cy), 3, cv::Scalar(255, 0, 0), 3);
      cv::imshow("cv_ptr->image", cv_ptr->image);
      cv::waitKey(1);
    }
  }

}  // namespace br2_tracking
```

br2_tracking/src/object_tracker_main.cpp

```cpp
#include <memory>

#include "br2_tracking/ObjectDetector.hpp"
#include "br2_tracking/HeadController.hpp"

#include "br2_tracking_msgs/msg/pan_tilt_command.hpp"

#include "lifecycle_msgs/msg/transition.hpp"
#include "rclcpp/rclcpp.hpp"

int main(int argc, char * argv[])
{
  rclcpp::init(argc, argv);

  auto node_detector = std::make_shared<br2_tracking::ObjectDetector>();
  auto node_head_controller = std::make_shared<br2_tracking::HeadController>();
  auto node_tracker = rclcpp::Node::make_shared("tracker");

  auto command_pub =
    node_tracker->create_publisher<br2_tracking_msgs::msg::PanTiltCommand>("/command", 100);
  auto detection_sub = node_tracker->create_subscription<vision_msgs::msg::Detection2D>(
    "/detection", rclcpp::SensorDataQoS(),
    [command_pub](vision_msgs::msg::Detection2D::SharedPtr msg) {
```

br2_tracking/src/object_tracker_main.cpp

```cpp
    br2_tracking_msgs::msg::PanTiltCommand command;
    command.pan = (msg->bbox.center.x / msg->source_img.width) * 2.0 - 1.0;
    command.tilt = (msg->bbox.center.y / msg->source_img.height) * 2.0 - 1.0;
    command_pub->publish(command);
  });

  rclcpp::executors::SingleThreadedExecutor executor;
  executor.add_node(node_detector);
  executor.add_node(node_head_controller->get_node_base_interface());
  executor.add_node(node_tracker);

  node_head_controller->trigger_transition(
    lifecycle_msgs::msg::Transition::TRANSITION_CONFIGURE);

  executor.spin();

  rclcpp::shutdown();
  return 0;
}
```

br2_tracking/src/object_detector_main.cpp

```cpp
#include <memory>

#include "br2_tracking/ObjectDetector.hpp"
#include "rclcpp/rclcpp.hpp"

int main(int argc, char * argv[])
{
  rclcpp::init(argc, argv);
  auto node_detector = std::make_shared<br2_tracking::ObjectDetector>();

  rclcpp::spin(node_detector);

  rclcpp::shutdown();
  return 0;
}
```

br2_tracking/tests/CMakeLists.txt

```
ament_add_gtest(pid_test pid_test.cpp)
ament_target_dependencies(pid_test ${dependencies})
target_link_libraries(pid_test ${PROJECT_NAME})
```

br2_tracking/tests/pid_test.cpp

```cpp
#include <random>

#include "br2_tracking/PIDController.hpp"

#include "gtest/gtest.h"

TEST(pid_tests, pid_test_1)
{
  br2_tracking::PIDController pid(0.0, 1.0, 0.0, 1.0);

  ASSERT_NEAR(pid.get_output(0.0), 0.0, 0.05);
  ASSERT_LT(pid.get_output(0.1), 0.099);
  ASSERT_GT(pid.get_output(0.1), -0.4);
  ASSERT_LT(pid.get_output(0.1), 0.3);
}

TEST(pid_tests, pid_test_2)
{
  br2_tracking::PIDController pid(0.0, 1.0, 0.0, 1.0);
```

```
br2_tracking/tests/pid_test.cpp
```

```cpp
  pid.set_pid(1.0, 0.0, 0.0);

  std::random_device rd;
  std::mt19937 gen(rd());
  std::uniform_real_distribution<> dis(-5.0, 5.0);

  for (int n = 0; n < 100000; n++) {
    double random_number = dis(gen);
    double output = pid.get_output(random_number);

    ASSERT_LE(output, 1.0);
    ASSERT_GE(output, -1.0);

    if (output < -2.0) {
      ASSERT_NEAR(output, -1.0, 0.01);
    }
    if (output > 2.0) {
      ASSERT_NEAR(output, 1.0, 0.01);
    }
    if (output > 0.0) {
      ASSERT_GT(output, 0.0);
    }
    if (output < 0.0) {
      ASSERT_LT(output, 0.0);
    }
  }
}

int main(int argc, char ** argv)
{
  testing::InitGoogleTest(&argc, argv);
  return RUN_ALL_TESTS();
}
```

A.8　BR2_BT_BUMPGO 软件包

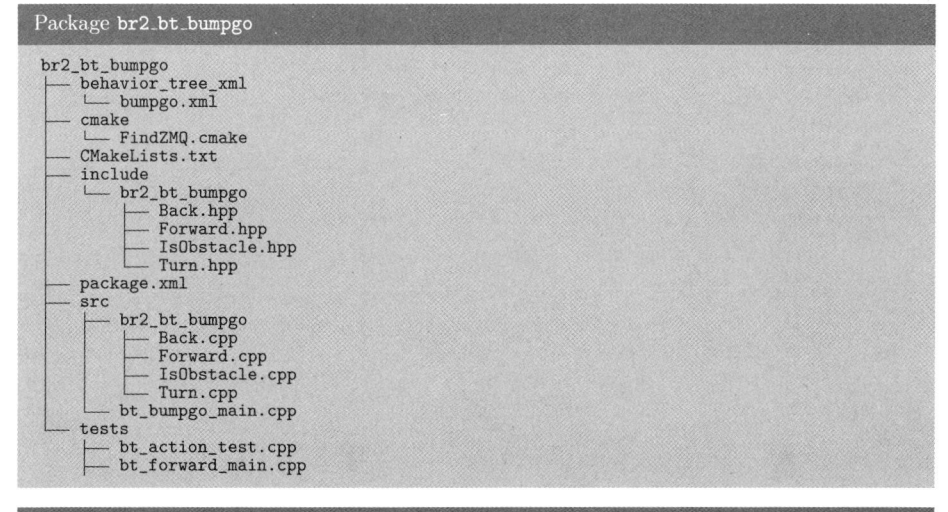

```
Package br2_bt_bumpgo
```

```
br2_bt_bumpgo
├── behavior_tree_xml
│   └── bumpgo.xml
├── cmake
│   └── FindZMQ.cmake
├── CMakeLists.txt
├── include
│   └── br2_bt_bumpgo
│       ├── Back.hpp
│       ├── Forward.hpp
│       ├── IsObstacle.hpp
│       └── Turn.hpp
├── package.xml
├── src
│   ├── br2_bt_bumpgo
│   │   ├── Back.cpp
│   │   ├── Forward.cpp
│   │   ├── IsObstacle.cpp
│   │   └── Turn.cpp
│   └── bt_bumpgo_main.cpp
└── tests
    ├── bt_action_test.cpp
    └── bt_forward_main.cpp
```

```
br2_bt_bumpgo/behavior_tree_xml/bumpgo.xml
```

```xml
<?xml version="1.0"?>
<root main_tree_to_execute="BehaviorTree">
```

```
br2_bt_bumpgo/behavior_tree_xml/bumpgo.xml
```

```xml
<!-- ////////// -->
<BehaviorTree ID="BehaviorTree">
    <ReactiveSequence>
        <Fallback>
            <Inverter>
                <Condition ID="IsObstacle" distance="1.0"/>
            </Inverter>
            <Sequence>
                <Action ID="Back"/>
                <Action ID="Turn"/>
            </Sequence>
        </Fallback>
        <Action ID="Forward"/>
    </ReactiveSequence>
</BehaviorTree>
<!-- ////////// -->
<TreeNodesModel>
    <Action ID="Back"/>
    <Action ID="Forward"/>
    <Condition ID="IsObstacle">
        <input_port default="1.0" name="distance">
          Distance to consider obstacle
        </input_port>
    </Condition>
    <Action ID="Turn"/>
</TreeNodesModel>
<!-- ////////// -->
</root>
```

```
br2_bt_bumpgo/CMakeLists.txt
```

```cmake
cmake_minimum_required(VERSION 3.5)
project(br2_bt_bumpgo)

set(CMAKE_CONFIG_PATH ${CMAKE_MODULE_PATH}  "${CMAKE_CURRENT_LIST_DIR}/cmake")
list(APPEND CMAKE_MODULE_PATH "${CMAKE_CONFIG_PATH}")

find_package(ament_cmake REQUIRED)
find_package(rclcpp REQUIRED)
find_package(behaviortree_cpp_v3 REQUIRED)
find_package(sensor_msgs REQUIRED)
find_package(geometry_msgs REQUIRED)
find_package(ament_index_cpp REQUIRED)

find_package(ZMQ)
if(ZMQ_FOUND)
    message(STATUS "ZeroMQ found.")
    add_definitions(-DZMQ_FOUND)
else()
  message(WARNING "ZeroMQ NOT found. Not including PublisherZMQ.")
endif()

set(CMAKE_CXX_STANDARD 17)

set(dependencies
    rclcpp
    behaviortree_cpp_v3
    sensor_msgs
    geometry_msgs
    ament_index_cpp
)

include_directories(include ${ZMQ_INCLUDE_DIRS})

add_library(br2_forward_bt_node SHARED src/br2_bt_bumpgo/Forward.cpp)
add_library(br2_back_bt_node SHARED src/br2_bt_bumpgo/Back.cpp)
add_library(br2_turn_bt_node SHARED src/br2_bt_bumpgo/Turn.cpp)
add_library(br2_is_obstacle_bt_node SHARED src/br2_bt_bumpgo/IsObstacle.cpp)
```

br2_bt_bumpgo/CMakeLists.txt

```cmake
list(APPEND plugin_libs
  br2_forward_bt_node
  br2_back_bt_node
  br2_turn_bt_node
  br2_is_obstacle_bt_node
)

foreach(bt_plugin ${plugin_libs})
  ament_target_dependencies(${bt_plugin} ${dependencies})
  target_compile_definitions(${bt_plugin} PRIVATE BT_PLUGIN_EXPORT)
endforeach()

add_executable(bt_bumpgo src/bt_bumpgo_main.cpp)
ament_target_dependencies(bt_bumpgo ${dependencies})
target_link_libraries(bt_bumpgo ${ZMQ_LIBRARIES})

install(TARGETS
  ${plugin_libs}
  bt_bumpgo
  ARCHIVE DESTINATION lib
  LIBRARY DESTINATION lib
  RUNTIME DESTINATION lib/${PROJECT_NAME}
)

install(DIRECTORY include/
  DESTINATION include/
)

install(DIRECTORY behavior_tree_xml
  DESTINATION share/${PROJECT_NAME}
)

if(BUILD_TESTING)
  find_package(ament_lint_auto REQUIRED)
  ament_lint_auto_find_test_dependencies()

  set(ament_cmake_cpplint_FOUND TRUE)
  ament_lint_auto_find_test_dependencies()

  find_package(ament_cmake_gtest REQUIRED)

  add_subdirectory(tests)
endif()

ament_export_include_directories(include)
ament_export_dependencies(${dependencies})

ament_package()
```

br2_bt_bumpgo/package.xml

```xml
<?xml version="1.0"?>
<?xml-model href="http://download.ros.org/schema/package_format3.xsd"
  schematypens="http://www.w3.org/2001/XMLSchema"?>
<package format="3">
  <name>br2_bt_bumpgo</name>
  <version>0.1.0</version>
  <description>Mastering Behavior Trees in ROS 2</description>
  <maintainer email="fmrico@gmail.com">Francisco Martǰn</maintainer>
  <license>Apache 2.0</license>

  <buildtool_depend>ament_cmake</buildtool_depend>

  <depend>rclcpp</depend>
  <depend>behaviortree_cpp_v3</depend>
  <depend>sensor_msgs</depend>
  <depend>geometry_msgs</depend>
  <depend>libzmq3-dev</depend>
  <depend>ament_index_cpp</depend>
```

br2_bt_bumpgo/package.xml

```xml
  <test_depend>ament_lint_auto</test_depend>
  <test_depend>ament_lint_common</test_depend>
  <test_depend>ament_cmake_gtest</test_depend>

  <export>
    <build_type>ament_cmake</build_type>
  </export>
</package>
```

br2_bt_bumpgo/include/br2_bt_bumpgo/Turn.hpp

```cpp
#ifndef BR2_BT_BUMPGO__TURN_HPP_
#define BR2_BT_BUMPGO__TURN_HPP_

#include <string>

#include "behaviortree_cpp_v3/behavior_tree.h"
#include "behaviortree_cpp_v3/bt_factory.h"

#include "geometry_msgs/msg/twist.hpp"
#include "rclcpp/rclcpp.hpp"

namespace br2_bt_bumpgo
{

class Turn : public BT::ActionNodeBase
{
public:
  explicit Turn(
    const std::string & xml_tag_name,
    const BT::NodeConfiguration & conf);

  void halt();
  BT::NodeStatus tick();

  static BT::PortsList providedPorts()
  {
    return BT::PortsList({});
  }

private:
  rclcpp::Node::SharedPtr node_;
  rclcpp::Time start_time_;
  rclcpp::Publisher<geometry_msgs::msg::Twist>::SharedPtr vel_pub_;
};

}  // namespace br2_bt_bumpgo

#endif  // BR2_BT_BUMPGO__TURN_HPP_
```

br2_bt_bumpgo/include/br2_bt_bumpgo/IsObstacle.hpp

```cpp
#ifndef BR2_BT_BUMPGO__ISOBSTACLE_HPP_
#define BR2_BT_BUMPGO__ISOBSTACLE_HPP_

#include <string>

#include "behaviortree_cpp_v3/behavior_tree.h"
#include "behaviortree_cpp_v3/bt_factory.h"

#include "sensor_msgs/msg/laser_scan.hpp"
#include "rclcpp/rclcpp.hpp"

namespace br2_bt_bumpgo
{

class IsObstacle : public BT::ConditionNode
{
```

br2_bt_bumpgo/include/br2_bt_bumpgo/IsObstacle.hpp

```cpp
public:
  explicit IsObstacle(
    const std::string & xml_tag_name,
    const BT::NodeConfiguration & conf);

  BT::NodeStatus tick();

  static BT::PortsList providedPorts()
  {
    return BT::PortsList(
      {
        BT::InputPort<double>("distance")
      });
  }

  void laser_callback(sensor_msgs::msg::LaserScan::UniquePtr msg);

private:
  rclcpp::Node::SharedPtr node_;
  rclcpp::Time last_reading_time_;
  rclcpp::Subscription<sensor_msgs::msg::LaserScan>::SharedPtr laser_sub_;
  sensor_msgs::msg::LaserScan::UniquePtr last_scan_;
};

}  // namespace br2_bt_bumpgo

#endif  // BR2_BT_BUMPGO__ISOBSTACLE_HPP_
```

br2_bt_bumpgo/include/br2_bt_bumpgo/Back.hpp

```cpp
#ifndef BR2_BT_BUMPGO__BACK_HPP_
#define BR2_BT_BUMPGO__BACK_HPP_

#include <string>

#include "behaviortree_cpp_v3/behavior_tree.h"
#include "behaviortree_cpp_v3/bt_factory.h"

#include "geometry_msgs/msg/twist.hpp"
#include "rclcpp/rclcpp.hpp"

namespace br2_bt_bumpgo
{

class Back : public BT::ActionNodeBase
{
public:
  explicit Back(
    const std::string & xml_tag_name,
    const BT::NodeConfiguration & conf);

  void halt();
  BT::NodeStatus tick();
```

br2_bt_bumpgo/include/br2_bt_bumpgo/Back.hpp

```cpp
  static BT::PortsList providedPorts()
  {
    return BT::PortsList({});
  }

private:
  rclcpp::Node::SharedPtr node_;
  rclcpp::Time start_time_;
  rclcpp::Publisher<geometry_msgs::msg::Twist>::SharedPtr vel_pub_;
};

}  // namespace br2_bt_bumpgo

#endif  // BR2_BT_BUMPGO__BACK_HPP_
```

```
br2_bt_bumpgo/include/br2_bt_bumpgo/Forward.hpp
```

```cpp
#ifndef BR2_BT_BUMPGO__FORWARD_HPP_
#define BR2_BT_BUMPGO__FORWARD_HPP_

#include <string>

#include "behaviortree_cpp_v3/behavior_tree.h"
#include "behaviortree_cpp_v3/bt_factory.h"

#include "geometry_msgs/msg/twist.hpp"
#include "rclcpp/rclcpp.hpp"

namespace br2_bt_bumpgo
{

class Forward : public BT::ActionNodeBase
{
public:
  explicit Forward(
    const std::string & xml_tag_name,
    const BT::NodeConfiguration & conf);

  void halt() {}
  BT::NodeStatus tick();

  static BT::PortsList providedPorts()
  {
    return BT::PortsList({});
  }

private:
  rclcpp::Node::SharedPtr node_;
  rclcpp::Publisher<geometry_msgs::msg::Twist>::SharedPtr vel_pub_;
};

}  // namespace br2_bt_bumpgo

#endif  // BR2_BT_BUMPGO__FORWARD_HPP_
```

```
br2_bt_bumpgo/src/bt_bumpgo_main.cpp
```

```cpp
#include <string>
#include <memory>

#include "behaviortree_cpp_v3/behavior_tree.h"
#include "behaviortree_cpp_v3/bt_factory.h"
#include "behaviortree_cpp_v3/utils/shared_library.h"
#include "behaviortree_cpp_v3/loggers/bt_zmq_publisher.h"

#include "ament_index_cpp/get_package_share_directory.hpp"

#include "rclcpp/rclcpp.hpp"

int main(int argc, char * argv[])
{
  rclcpp::init(argc, argv);

  auto node = rclcpp::Node::make_shared("patrolling_node");

  BT::BehaviorTreeFactory factory;
  BT::SharedLibrary loader;

  factory.registerFromPlugin(loader.getOSName("br2_forward_bt_node"));
  factory.registerFromPlugin(loader.getOSName("br2_back_bt_node"));
  factory.registerFromPlugin(loader.getOSName("br2_turn_bt_node"));
  factory.registerFromPlugin(loader.getOSName("br2_is_obstacle_bt_node"));

  std::string pkgpath = ament_index_cpp::get_package_share_directory("br2_bt_bumpgo");
  std::string xml_file = pkgpath + "/behavior_tree_xml/bumpgo.xml";
```

br2_bt_bumpgo/src/bt_bumpgo_main.cpp

```cpp
auto blackboard = BT::Blackboard::create();
blackboard->set("node", node);
BT::Tree tree = factory.createTreeFromFile(xml_file, blackboard);

auto publisher_zmq = std::make_shared<BT::PublisherZMQ>(tree, 10, 1666, 1667);

rclcpp::Rate rate(10);

bool finish = false;
while (!finish && rclcpp::ok()) {
  finish = tree.rootNode()->executeTick() != BT::NodeStatus::RUNNING;

  rclcpp::spin_some(node);
  rate.sleep();
}

rclcpp::shutdown();
return 0;
}
```

br2_bt_bumpgo/src/br2_bt_bumpgo/Back.cpp

```cpp
#include <string>
#include <iostream>

#include "br2_bt_bumpgo/Back.hpp"

#include "behaviortree_cpp_v3/behavior_tree.h"

#include "geometry_msgs/msg/twist.hpp"
#include "rclcpp/rclcpp.hpp"

namespace br2_bt_bumpgo
{

using namespace std::chrono_literals;

Back::Back(
  const std::string & xml_tag_name,
  const BT::NodeConfiguration & conf)
: BT::ActionNodeBase(xml_tag_name, conf)
{
  config().blackboard->get("node", node_);

  vel_pub_ = node_->create_publisher<geometry_msgs::msg::Twist>("/output_vel", 100);
}

void
Back::halt()
{
}

BT::NodeStatus
Back::tick()
{
  if (status() == BT::NodeStatus::IDLE) {
    start_time_ = node_->now();
  }

  geometry_msgs::msg::Twist vel_msgs;
  vel_msgs.linear.x = -0.3;
  vel_pub_->publish(vel_msgs);

  auto elapsed = node_->now() - start_time_;

  if (elapsed < 3s) {
    return BT::NodeStatus::RUNNING;
  } else {
    return BT::NodeStatus::SUCCESS;
```

br2_bt_bumpgo/src/br2_bt_bumpgo/Back.cpp

```cpp
  }
}

}  // namespace br2_bt_bumpgo

#include "behaviortree_cpp_v3/bt_factory.h"
BT_REGISTER_NODES(factory)
{
  factory.registerNodeType<br2_bt_bumpgo::Back>("Back");
}
```

br2_bt_bumpgo/src/br2_bt_bumpgo/Forward.cpp

```cpp
#include <string>
#include <iostream>

#include "br2_bt_bumpgo/Forward.hpp"

#include "behaviortree_cpp_v3/behavior_tree.h"

#include "geometry_msgs/msg/twist.hpp"
#include "rclcpp/rclcpp.hpp"

namespace br2_bt_bumpgo
{

using namespace std::chrono_literals;

Forward::Forward(
  const std::string & xml_tag_name,
  const BT::NodeConfiguration & conf)
: BT::ActionNodeBase(xml_tag_name, conf)
{
  config().blackboard->get("node", node_);

  vel_pub_ = node_->create_publisher<geometry_msgs::msg::Twist>("/output_vel", 100);
}

BT::NodeStatus
Forward::tick()
{
  geometry_msgs::msg::Twist vel_msgs;
  vel_msgs.linear.x = 0.3;
  vel_pub_->publish(vel_msgs);

  return BT::NodeStatus::RUNNING;
}

}  // namespace br2_bt_bumpgo

#include "behaviortree_cpp_v3/bt_factory.h"
BT_REGISTER_NODES(factory)
{
  factory.registerNodeType<br2_bt_bumpgo::Forward>("Forward");
}
```

br2_bt_bumpgo/src/br2_bt_bumpgo/IsObstacle.cpp

```cpp
#include <string>
#include <utility>

#include "br2_bt_bumpgo/IsObstacle.hpp"

#include "behaviortree_cpp_v3/behavior_tree.h"

#include "sensor_msgs/msg/laser_scan.hpp"
#include "rclcpp/rclcpp.hpp"
```

br2_bt_bumpgo/src/br2_bt_bumpgo/IsObstacle.cpp

```cpp
namespace br2_bt_bumpgo
{

using namespace std::chrono_literals;
using namespace std::placeholders;

IsObstacle::IsObstacle(
  const std::string & xml_tag_name,
  const BT::NodeConfiguration & conf)
: BT::ConditionNode(xml_tag_name, conf)
{
  config().blackboard->get("node", node_);

  laser_sub_ = node_->create_subscription<sensor_msgs::msg::LaserScan>(
    "/input_scan", 100, std::bind(&IsObstacle::laser_callback, this, _1));

  last_reading_time_ = node_->now();
}

void
IsObstacle::laser_callback(sensor_msgs::msg::LaserScan::UniquePtr msg)
{
  last_scan_ = std::move(msg);
}

BT::NodeStatus
IsObstacle::tick()
{
  if (last_scan_ == nullptr) {
    return BT::NodeStatus::FAILURE;
  }

  double distance = 1.0;
  getInput("distance", distance);

  if (last_scan_->ranges[last_scan_->ranges.size() / 2] < distance) {
    return BT::NodeStatus::SUCCESS;
  } else {
    return BT::NodeStatus::FAILURE;
  }
}

}  // namespace br2_bt_bumpgo

#include "behaviortree_cpp_v3/bt_factory.h"
BT_REGISTER_NODES(factory)
{
  factory.registerNodeType<br2_bt_bumpgo::IsObstacle>("IsObstacle");
}
```

br2_bt_bumpgo/src/br2_bt_bumpgo/Turn.cpp

```cpp
#include <string>
#include <iostream>

#include "br2_bt_bumpgo/Turn.hpp"

#include "behaviortree_cpp_v3/behavior_tree.h"

#include "geometry_msgs/msg/twist.hpp"
#include "rclcpp/rclcpp.hpp"

namespace br2_bt_bumpgo
{

using namespace std::chrono_literals;

Turn::Turn(
```

```
br2_bt_bumpgo/src/br2_bt_bumpgo/Turn.cpp
```

```cpp
  const std::string & xml_tag_name,
  const BT::NodeConfiguration & conf)
: BT::ActionNodeBase(xml_tag_name, conf)
{
  config().blackboard->get("node", node_);

  vel_pub_ = node_->create_publisher<geometry_msgs::msg::Twist>("/output_vel", 100);
}

void
Turn::halt()
{
}

BT::NodeStatus
Turn::tick()
{
  if (status() == BT::NodeStatus::IDLE) {
    start_time_ = node_->now();
  }

  geometry_msgs::msg::Twist vel_msgs;
  vel_msgs.angular.z = 0.5;
  vel_pub_->publish(vel_msgs);

  auto elapsed = node_->now() - start_time_;

  if (elapsed < 3s) {
    return BT::NodeStatus::RUNNING;
  } else {
    return BT::NodeStatus::SUCCESS;
  }
}

}  // namespace br2_bt_bumpgo

#include "behaviortree_cpp_v3/bt_factory.h"
BT_REGISTER_NODES(factory)
{
  factory.registerNodeType<br2_bt_bumpgo::Turn>("Turn");
}
```

```
br2_bt_bumpgo/tests/bt_action_test.cpp
```

```cpp
#include <string>
#include <list>
#include <memory>
#include <vector>
#include <set>

#include "behaviortree_cpp_v3/behavior_tree.h"
#include "behaviortree_cpp_v3/bt_factory.h"
#include "behaviortree_cpp_v3/utils/shared_library.h"

#include "ament_index_cpp/get_package_share_directory.hpp"

#include "geometry_msgs/msg/twist.hpp"
#include "sensor_msgs/msg/laser_scan.hpp"

#include "rclcpp/rclcpp.hpp"
#include "rclcpp_action/rclcpp_action.hpp"

#include "gtest/gtest.h"

using namespace std::placeholders;
using namespace std::chrono_literals;


class VelocitySinkNode : public rclcpp::Node
```

```
br2_bt_bumpgo/tests/bt_action_test.cpp
{
public:
  VelocitySinkNode()
  : Node("VelocitySink")
  {
    vel_sub_ = create_subscription<geometry_msgs::msg::Twist>(
      "/output_vel", 100, std::bind(&VelocitySinkNode::vel_callback, this, _1));
  }

  void vel_callback(geometry_msgs::msg::Twist::SharedPtr msg)
  {
    vel_msgs_.push_back(*msg);
  }

  std::list<geometry_msgs::msg::Twist> vel_msgs_;
private:
  rclcpp::Subscription<geometry_msgs::msg::Twist>::SharedPtr vel_sub_;
};

TEST(bt_action, turn_btn)
{
  auto node = rclcpp::Node::make_shared("turn_btn_node");
  auto node_sink = std::make_shared<VelocitySinkNode>();

  BT::BehaviorTreeFactory factory;
  BT::SharedLibrary loader;

  factory.registerFromPlugin(loader.getOSName("br2_turn_bt_node"));

  std::string xml_bt =
    R"(
    <root main_tree_to_execute = "MainTree" >
      <BehaviorTree ID="MainTree">
          <Turn />
      </BehaviorTree>
    </root>)";

  auto blackboard = BT::Blackboard::create();
  blackboard->set("node", node);
  BT::Tree tree = factory.createTreeFromText(xml_bt, blackboard);

  rclcpp::Rate rate(10);
  bool finish = false;
  while (!finish && rclcpp::ok()) {
    finish = tree.rootNode()->executeTick() == BT::NodeStatus::SUCCESS;
    rclcpp::spin_some(node_sink);
    rate.sleep();
  }

  ASSERT_FALSE(node_sink->vel_msgs_.empty());
  ASSERT_NEAR(node_sink->vel_msgs_.size(), 30, 1);

  geometry_msgs::msg::Twist & one_twist = node_sink->vel_msgs_.front();

  ASSERT_GT(one_twist.angular.z, 0.1);
  ASSERT_NEAR(one_twist.linear.x, 0.0, 0.0000001);
}

TEST(bt_action, back_btn)
{
  auto node = rclcpp::Node::make_shared("back_btn_node");
  auto node_sink = std::make_shared<VelocitySinkNode>();

  BT::BehaviorTreeFactory factory;
  BT::SharedLibrary loader;

  factory.registerFromPlugin(loader.getOSName("br2_back_bt_node"));

  std::string xml_bt =gte_node
```

```cpp
br2_bt_bumpgo/tests/bt_action_test.cpp
    R"(
    <root main_tree_to_execute = "MainTree" >
      <BehaviorTree ID="MainTree">
          <Back />
      </BehaviorTree>
    </root>)";

  auto blackboard = BT::Blackboard::create();
  blackboard->set("node", node);
  BT::Tree tree = factory.createTreeFromText(xml_bt, blackboard);

  rclcpp::Rate rate(10);
  bool finish = false;
  while (!finish && rclcpp::ok()) {
    finish = tree.rootNode()->executeTick() == BT::NodeStatus::SUCCESS;
    rclcpp::spin_some(node_sink);
    rate.sleep();
  }

  ASSERT_FALSE(node_sink->vel_msgs_.empty());
  ASSERT_NEAR(node_sink->vel_msgs_.size(), 30, 1);

  geometry_msgs::msg::Twist & one_twist = node_sink->vel_msgs_.front();

  ASSERT_LT(one_twist.linear.x, -0.1);
  ASSERT_NEAR(one_twist.angular.z, 0.0, 0.0000001);
}

TEST(bt_action, forward_btn)
{
  auto node = rclcpp::Node::make_shared("forward_btn_node");
  auto node_sink = std::make_shared<VelocitySinkNode>();

  BT::BehaviorTreeFactory factory;
  BT::SharedLibrary loader;

  factory.registerFromPlugin(loader.getOSName("br2_forward_bt_node"));

  std::string xml_bt =
    R"(
    <root main_tree_to_execute = "MainTree" >
      <BehaviorTree ID="MainTree">
          <Forward />
      </BehaviorTree>
    </root>)";

  auto blackboard = BT::Blackboard::create();
  blackboard->set("node", node);
  BT::Tree tree = factory.createTreeFromText(xml_bt, blackboard);

  rclcpp::Rate rate(10);
  auto current_status = BT::NodeStatus::FAILURE;
  int counter = 0;
  while (counter++ < 30 && rclcpp::ok()) {
    current_status = tree.rootNode()->executeTick();
    rclcpp::spin_some(node_sink);
    rate.sleep();
  }

  ASSERT_EQ(current_status, BT::NodeStatus::RUNNING);
  ASSERT_FALSE(node_sink->vel_msgs_.empty());
  ASSERT_NEAR(node_sink->vel_msgs_.size(), 30, 1);

  geometry_msgs::msg::Twist & one_twist = node_sink->vel_msgs_.front();

  ASSERT_GT(one_twist.linear.x, 0.1);
  ASSERT_NEAR(one_twist.angular.z, 0.0, 0.0000001);
}

TEST(bt_action, is_obstacle_btn)
{
```

br2_bt_bumpgo/tests/bt_action_test.cpp

```cpp
  auto node = rclcpp::Node::make_shared("is_obstacle_btn_node");
  auto scan_pub = node->create_publisher<sensor_msgs::msg::LaserScan>("input_scan", 1);

  BT::BehaviorTreeFactory factory;
  BT::SharedLibrary loader;

  factory.registerFromPlugin(loader.getOSName("br2_is_obstacle_bt_node"));

  std::string xml_bt =
    R"(
    <root main_tree_to_execute = "MainTree" >
      <BehaviorTree ID="MainTree">
        <IsObstacle/>
      </BehaviorTree>
    </root>)";
  auto blackboard = BT::Blackboard::create();
  blackboard->set("node", node);
  BT::Tree tree = factory.createTreeFromText(xml_bt, blackboard);

  rclcpp::Rate rate(10);

  sensor_msgs::msg::LaserScan scan;
  scan.ranges.push_back(2.0);
  for (int i = 0; i < 10; i++) {
    scan_pub->publish(scan);
    rclcpp::spin_some(node);
    rate.sleep();
  }

  BT::NodeStatus current_status = tree.rootNode()->executeTick();
  ASSERT_EQ(current_status, BT::NodeStatus::FAILURE);

  scan.ranges[0] = 0.3;
  for (int i = 0; i < 10; i++) {
    scan_pub->publish(scan);
    rclcpp::spin_some(node);
    rate.sleep();
  }

  current_status = tree.rootNode()->executeTick();
  ASSERT_EQ(current_status, BT::NodeStatus::SUCCESS);

  xml_bt =
    R"(
    <root main_tree_to_execute = "MainTree" >
      <BehaviorTree ID="MainTree">
        <IsObstacle distance="0.5"/>
      </BehaviorTree>
    </root>)";
  tree = factory.createTreeFromText(xml_bt, blackboard);

  scan.ranges[0] = 0.3;
  for (int i = 0; i < 10; i++) {
    scan_pub->publish(scan);
    rclcpp::spin_some(node);
    rate.sleep();
  }

  current_status = tree.rootNode()->executeTick();
  ASSERT_EQ(current_status, BT::NodeStatus::SUCCESS);

  scan.ranges[0] = 0.6;
  for (int i = 0; i < 10; i++) {
    scan_pub->publish(scan);
    rclcpp::spin_some(node);
    rate.sleep();
  }

  current_status = tree.rootNode()->executeTick();
  ASSERT_EQ(current_status, BT::NodeStatus::FAILURE);
```

br2_bt_bumpgo/tests/bt_action_test.cpp

```cpp
}

int main(int argc, char ** argv)
{
  rclcpp::init(argc, argv);

  testing::InitGoogleTest(&argc, argv);
  return RUN_ALL_TESTS();
}
```

br2_bt_bumpgo/tests/CMakeLists.txt

```
ament_add_gtest(bt_action_test bt_action_test.cpp)
ament_target_dependencies(bt_action_test ${dependencies})

add_executable(bt_forward bt_forward_main.cpp)
ament_target_dependencies(bt_forward ${dependencies})
target_link_libraries(bt_forward ${ZMQ_LIBRARIES})

install(TARGETS
  bt_forward
  ARCHIVE DESTINATION lib
  LIBRARY DESTINATION lib
  RUNTIME DESTINATION lib/${PROJECT_NAME}
)
```

br2_bt_bumpgo/tests/bt_forward_main.cpp

```cpp
#include <string>
#include <memory>

#include "behaviortree_cpp_v3/behavior_tree.h"
#include "behaviortree_cpp_v3/bt_factory.h"
#include "behaviortree_cpp_v3/utils/shared_library.h"
#include "behaviortree_cpp_v3/loggers/bt_zmq_publisher.h"

#include "ament_index_cpp/get_package_share_directory.hpp"

#include "rclcpp/rclcpp.hpp"

int main(int argc, char * argv[])
{
  rclcpp::init(argc, argv);

  auto node = rclcpp::Node::make_shared("forward_node");

  BT::BehaviorTreeFactory factory;
  BT::SharedLibrary loader;

  factory.registerFromPlugin(loader.getOSName("br2_forward_bt_node"));

  std::string xml_bt =
    R"(
  <root main_tree_to_execute = "MainTree" >
    <BehaviorTree ID="MainTree">
        <Forward />
    </BehaviorTree>
  </root>)";

  auto blackboard = BT::Blackboard::create();
  blackboard->set("node", node);
  BT::Tree tree = factory.createTreeFromText(xml_bt, blackboard);

  rclcpp::Rate rate(10);
  bool finish = false;
  while (!finish && rclcpp::ok()) {
```

```
br2_bt_bumpgo/tests/bt_forward_main.cpp
    finish = tree.rootNode()->executeTick() != BT::NodeStatus::RUNNING;

    rclcpp::spin_some(node);
    rate.sleep();
  }

  rclcpp::shutdown();
  return 0;
}
```

A.9 BR2_BT_PATROLLING 软件包

```
Package br2_bt_patrolling
br2_bt_patrolling
├── behavior_tree_xml
│   └── patrolling.xml
├── cmake
│   └── FindZMQ.cmake
├── CMakeLists.txt
├── include
│   └── br2_bt_patrolling
│       ├── BatteryChecker.hpp
│       ├── ctrl_support
│       │   ├── BTActionNode.hpp
│       │   └── BTLifecycleCtrlNode.hpp
│       ├── GetWaypoint.hpp
│       ├── Move.hpp
│       ├── Patrol.hpp
│       ├── Recharge.hpp
│       └── TrackObjects.hpp
├── launch
│   └── patrolling.launch.py
├── package.xml
├── src
│   ├── br2_bt_patrolling
│   │   ├── BatteryChecker.cpp
│   │   ├── GetWaypoint.cpp
│   │   ├── Move.cpp
│   │   ├── Patrol.cpp
│   │   ├── Recharge.cpp
│   │   └── TrackObjects.cpp
│   └── patrolling_main.cpp
└── tests
    ├── bt_action_test.cpp
    └── CMakeLists.txt
```

```
br2_bt_patrolling/behavior_tree_xml/patrolling.xml
<?xml version="1.0"?>
<root main_tree_to_execute="BehaviorTree">
    <!-- ////////// -->
    <BehaviorTree ID="BehaviorTree">
        <KeepRunningUntilFailure>
            <ReactiveSequence>
                <Fallback>
                    <Action ID="BatteryChecker"/>
                    <Sequence>
                        <Action ID="GetWaypoint" waypoint="{recharge_wp}" wp_id="recharge"/>
                        <Action ID="Move" goal="{recharge_wp}"/>
                        <Action ID="Recharge"/>
                    </Sequence>
                </Fallback>
                <Sequence>
```

br2_bt_patrolling/behavior_tree_xml/patrolling.xml

```xml
                    <Action ID="GetWaypoint" waypoint="{wp}" wp_id="next"/>
                    <Parallel success_threshold="1" failure_threshold="1">
                        <Action ID="TrackObjects"/>
                        <Action ID="Move" goal="{wp}"/>
                    </Parallel>
                    <Action ID="Patrol"/>
                </Sequence>
            </ReactiveSequence>
        </KeepRunningUntilFailure>
    </BehaviorTree>
    <!-- ////////// -->
    <TreeNodesModel>
        <Action ID="BatteryChecker"/>
        <Action ID="GetWaypoint">
            <output_port name="waypoint"/>
            <input_port name="wp_id"/>
        </Action>
        <Action ID="Move">
            <input_port name="goal"/>
        </Action>
        <Action ID="Patrol"/>
        <Action ID="Recharge"/>
        <Action ID="TrackObjects"/>
    </TreeNodesModel>
    <!-- ////////// -->
</root>
```

br2_bt_patrolling/CMakeLists.txt

```
cmake_minimum_required(VERSION 3.5)
project(br2_bt_patrolling)

set(CMAKE_BUILD_TYPE Debug)

set(CMAKE_CONFIG_PATH ${CMAKE_MODULE_PATH}  "${CMAKE_CURRENT_LIST_DIR}/cmake")
list(APPEND CMAKE_MODULE_PATH "${CMAKE_CONFIG_PATH}")

find_package(ament_cmake REQUIRED)
find_package(rclcpp REQUIRED)
find_package(rclcpp_lifecycle REQUIRED)
find_package(rclcpp_action REQUIRED)
find_package(behaviortree_cpp_v3 REQUIRED)
find_package(action_msgs REQUIRED)
find_package(lifecycle_msgs REQUIRED)
find_package(geometry_msgs REQUIRED)
find_package(nav2_msgs REQUIRED)
find_package(ament_index_cpp REQUIRED)

find_package(ZMQ)
if(ZMQ_FOUND)
    message(STATUS "ZeroMQ found.")
    add_definitions(-DZMQ_FOUND)
else()
  message(WARNING "ZeroMQ NOT found. Not including PublisherZMQ.")
endif()

set(CMAKE_CXX_STANDARD 17)

set(dependencies
    rclcpp
    rclcpp_lifecycle
    rclcpp_action
    behaviortree_cpp_v3
    action_msgs
    lifecycle_msgs
    geometry_msgs
    nav2_msgs
    ament_index_cpp
)

include_directories(include ${ZMQ_INCLUDE_DIRS})
```

br2_bt_patrolling/CMakeLists.txt

```cmake
add_library(br2_recharge_bt_node SHARED src/br2_bt_patrolling/Recharge.cpp)
add_library(br2_patrol_bt_node SHARED src/br2_bt_patrolling/Patrol.cpp)
add_library(br2_move_bt_node SHARED src/br2_bt_patrolling/Move.cpp)
add_library(br2_get_waypoint_bt_node SHARED src/br2_bt_patrolling/GetWaypoint.cpp)
add_library(br2_battery_checker_bt_node SHARED src/br2_bt_patrolling/BatteryChecker.cpp)
add_library(br2_track_objects_bt_node SHARED src/br2_bt_patrolling/TrackObjects.cpp)
list(APPEND plugin_libs
  br2_recharge_bt_node
  br2_patrol_bt_node
  br2_move_bt_node
  br2_get_waypoint_bt_node
  br2_battery_checker_bt_node
  br2_track_objects_bt_node
)

foreach(bt_plugin ${plugin_libs})
  ament_target_dependencies(${bt_plugin} ${dependencies})
  target_compile_definitions(${bt_plugin} PRIVATE BT_PLUGIN_EXPORT)
endforeach()

add_executable(patrolling_main src/patrolling_main.cpp)
ament_target_dependencies(patrolling_main ${dependencies})
target_link_libraries(patrolling_main ${ZMQ_LIBRARIES})

install(TARGETS
  ${plugin_libs}
  patrolling_main
  ARCHIVE DESTINATION lib
  LIBRARY DESTINATION lib
  RUNTIME DESTINATION lib/${PROJECT_NAME}
)

install(DIRECTORY include/
  DESTINATION include/
)

install(DIRECTORY behavior_tree_xml launch
  DESTINATION share/${PROJECT_NAME}
)


if(BUILD_TESTING)
  find_package(ament_lint_auto REQUIRED)
  ament_lint_auto_find_test_dependencies()

  set(ament_cmake_cpplint_FOUND TRUE)
  ament_lint_auto_find_test_dependencies()

  find_package(ament_cmake_gtest REQUIRED)

  add_subdirectory(tests)
endif()

ament_export_include_directories(include)
ament_export_dependencies(${dependencies})

ament_package()
```

br2_bt_patrolling/launch/patrolling.launch.py

```python
import os

from ament_index_python.packages import get_package_share_directory

from launch import LaunchDescription
from launch.actions import IncludeLaunchDescription
from launch.launch_description_sources import PythonLaunchDescriptionSource
from launch_ros.actions import Node
```

br2_bt_patrolling/launch/patrolling.launch.py

```python
def generate_launch_description():

    tracking_dir = get_package_share_directory('br2_tracking')

    tracking_cmd = IncludeLaunchDescription(
        PythonLaunchDescriptionSource(
          os.path.join(tracking_dir, 'launch', 'tracking.launch.py')))

    patrolling_cmd = Node(
        package='br2_bt_patrolling',
        executable='patrolling_main',
        parameters=[{
          'use_sim_time': True
        }],
        remappings=[
          ('input_scan', '/scan_raw'),
          ('output_vel', '/nav_vel'),
        ],
        output='screen'
    )

    ld = LaunchDescription()

    # Add any actions
    ld.add_action(tracking_cmd)
    ld.add_action(patrolling_cmd)

    return ld
```

br2_bt_patrolling/package.xml

```xml
<?xml version="1.0"?>
<?xml-model href="http://download.ros.org/schema/package_format3.xsd"
  schematypens="http://www.w3.org/2001/XMLSchema"?>
<package format="3">
  <name>br2_bt_patrolling</name>
  <version>0.1.0</version>
  <description>Patrolling behavior package</description>
  <maintainer email="fmrico@gmail.com">Francisco Martín</maintainer>
  <license>Apache 2.0</license>

  <buildtool_depend>ament_cmake</buildtool_depend>

  <depend>rclcpp</depend>
  <depend>rclcpp_lifecycle</depend>
  <depend>rclcpp_action</depend>
  <depend>behaviortree_cpp_v3</depend>
  <depend>action_msgs</depend>
  <depend>geometry_msgs</depend>
  <depend>lifecycle_msgs</depend>
  <depend>nav2_msgs</depend>
  <depend>libzmq3-dev</depend>
  <depend>ament_index_cpp</depend>

  <test_depend>ament_lint_auto</test_depend>
  <test_depend>ament_lint_common</test_depend>
  <test_depend>ament_cmake_gtest</test_depend>

  <export>
    <build_type>ament_cmake</build_type>
  </export>
</package>
```

br2_bt_patrolling/include/br2_bt_patrolling/BatteryChecker.hpp

```cpp
#ifndef BR2_BT_PATROLLING__BATTERYCHECKER_HPP_
#define BR2_BT_PATROLLING__BATTERYCHECKER_HPP_
```

br2_bt_patrolling/include/br2_bt_patrolling/BatteryChecker.hpp

```cpp
#include <string>
#include <vector>

#include "behaviortree_cpp_v3/behavior_tree.h"
#include "behaviortree_cpp_v3/bt_factory.h"

#include "geometry_msgs/msg/twist.hpp"

#include "rclcpp/rclcpp.hpp"

namespace br2_bt_patrolling
{

class BatteryChecker : public BT::ConditionNode
{
public:
  explicit BatteryChecker(
    const std::string & xml_tag_name,
    const BT::NodeConfiguration & conf);

  BT::NodeStatus tick();

  static BT::PortsList providedPorts()
  {
    return BT::PortsList({});
  }

  void vel_callback(const geometry_msgs::msg::Twist::SharedPtr msg);

  const float DECAY_LEVEL = 0.5;   // 0.5 * |vel| * dt
  const float EPSILON = 0.01;      // 0.001 * dt
  const float MIN_LEVEL = 10.0;
private:
  void update_battery();

  rclcpp::Node::SharedPtr node_;
  rclcpp::Time last_reading_time_;
  geometry_msgs::msg::Twist last_twist_;
  rclcpp::Subscription<geometry_msgs::msg::Twist>::SharedPtr vel_sub_;
};

}  // namespace br2_bt_patrolling

#endif  // BR2_BT_PATROLLING__BATTERYCHECKER_HPP_
```

br2_bt_patrolling/include/br2_bt_patrolling/ctrl_support/BTLifecycleCtrlNode.hpp

```cpp
#ifndef BR2_BT_PATROLLING__CTRL_SUPPORT__BTLIFECYCLECTRLNODE_HPP_
#define BR2_BT_PATROLLING__CTRL_SUPPORT__BTLIFECYCLECTRLNODE_HPP_

#include <memory>
#include <string>

#include "lifecycle_msgs/srv/change_state.hpp"
#include "lifecycle_msgs/srv/get_state.hpp"
#include "lifecycle_msgs/msg/state.hpp"

#include "behaviortree_cpp_v3/action_node.h"
#include "rclcpp/rclcpp.hpp"

namespace br2_bt_patrolling
{

using namespace std::chrono_literals;  // NOLINT

class BtLifecycleCtrlNode : public BT::ActionNodeBase
{
```

```cpp
br2_bt_patrolling/include/br2_bt_patrolling/ctrl_support/BTLifecycleCtrlNode.hpp
public:
  BtLifecycleCtrlNode(
    const std::string & xml_tag_name,
    const std::string & node_name,
    const BT::NodeConfiguration & conf)
  : BT::ActionNodeBase(xml_tag_name, conf), ctrl_node_name_(node_name)
  {
    node_ = config().blackboard->get<rclcpp::Node::SharedPtr>("node");
  }

  BtLifecycleCtrlNode() = delete;

  virtual ~BtLifecycleCtrlNode()
  {
  }

  template<typename serviceT>
  typename rclcpp::Client<serviceT>::SharedPtr createServiceClient(
    const std::string & service_name)
  {
    auto srv = node_->create_client<serviceT>(service_name);
    while (!srv->wait_for_service(1s)) {
      if (!rclcpp::ok()) {
        RCLCPP_ERROR(node_->get_logger(), "Interrupted while waiting for the service.");
      } else {
        RCLCPP_INFO(node_->get_logger(), "service not available, waiting again...");
      }
    }
    return srv;
  }

  virtual void on_tick() {}

  virtual BT::NodeStatus on_success()
  {
    return BT::NodeStatus::SUCCESS;
  }

  virtual BT::NodeStatus on_failure()
  {
    return BT::NodeStatus::FAILURE;
  }

  BT::NodeStatus tick() override
  {
    if (status() == BT::NodeStatus::IDLE) {
      change_state_client_ = createServiceClient<lifecycle_msgs::srv::ChangeState>(
        ctrl_node_name_ + "/change_state");
      get_state_client_ = createServiceClient<lifecycle_msgs::srv::GetState>(
        ctrl_node_name_ + "/get_state");
    }

    if (ctrl_node_state_ != lifecycle_msgs::msg::State::PRIMARY_STATE_ACTIVE) {
      ctrl_node_state_ = get_state();
      set_state(lifecycle_msgs::msg::State::PRIMARY_STATE_ACTIVE);
    }

    on_tick();

    return BT::NodeStatus::RUNNING;
  }

  void halt() override
  {
    if (ctrl_node_state_ == lifecycle_msgs::msg::State::PRIMARY_STATE_ACTIVE) {
      set_state(lifecycle_msgs::msg::State::PRIMARY_STATE_INACTIVE);
    }
    setStatus(BT::NodeStatus::IDLE);
  }

  // Get the state of the controlled node
  uint8_t get_state()
```

br2_bt_patrolling/include/br2_bt_patrolling/ctrl_support/BTLifecycleCtrlNode.hpp

```cpp
{
  auto request = std::make_shared<lifecycle_msgs::srv::GetState::Request>();
  auto result = get_state_client_->async_send_request(request);

  if (rclcpp::spin_until_future_complete(node_, result) !=
    rclcpp::FutureReturnCode::SUCCESS)
  {
    lifecycle_msgs::msg::State get_state;

    RCLCPP_ERROR(node_->get_logger(), "Failed to call get_state service");
    return lifecycle_msgs::msg::State::PRIMARY_STATE_UNKNOWN;
  }

  return result.get()->current_state.id;
}
// Get the state of the controlled node. Ot can fail, if not transition possible
bool set_state(uint8_t state)
{
  auto request = std::make_shared<lifecycle_msgs::srv::ChangeState::Request>();

  if (state == lifecycle_msgs::msg::State::PRIMARY_STATE_ACTIVE &&
    ctrl_node_state_ == lifecycle_msgs::msg::State::PRIMARY_STATE_INACTIVE)
  {
    request->transition.id = lifecycle_msgs::msg::Transition::TRANSITION_ACTIVATE;
  } else {
    if (state == lifecycle_msgs::msg::State::PRIMARY_STATE_INACTIVE &&
      ctrl_node_state_ == lifecycle_msgs::msg::State::PRIMARY_STATE_ACTIVE)
    {
      request->transition.id = lifecycle_msgs::msg::Transition::TRANSITION_DEACTIVATE;
    } else {
      if (state != ctrl_node_state_) {
        RCLCPP_ERROR(
          node_->get_logger(),
          "Transition not possible %u -> %u", ctrl_node_state_, state);
        return false;
      } else {
        return true;
      }
    }
  }

  auto result = change_state_client_->async_send_request(request);

  if (rclcpp::spin_until_future_complete(node_, result) !=
    rclcpp::FutureReturnCode::SUCCESS)
  {
    RCLCPP_ERROR(node_->get_logger(), "Failed to call set_state service");
    return false;
  }

  if (!result.get()->success) {
    RCLCPP_ERROR(
      node_->get_logger(),
      "Failed to set node state %u -> %u", ctrl_node_state_, state);
    return false;
  } else {
    RCLCPP_INFO(
      node_->get_logger(), "Transition success  %u -> %u", ctrl_node_state_, state);
  }

  ctrl_node_state_ = state;
  return true;
}

std::string ctrl_node_name_;
uint8_t ctrl_node_state_;

rclcpp::Client<lifecycle_msgs::srv::ChangeState>::SharedPtr change_state_client_;
rclcpp::Client<lifecycle_msgs::srv::GetState>::SharedPtr get_state_client_;
```

```
br2_bt_patrolling/include/br2_bt_patrolling/ctrl_support/BTLifecycleCtrlNode.hpp
```
```cpp
  rclcpp::Node::SharedPtr node_;
};


}  // namespace br2_bt_patrolling

#endif  // BR2_BT_PATROLLING__CTRL_SUPPORT__BTLIFECYCLECTRLNODE_HPP_
```

```
br2_bt_patrolling/include/br2_bt_patrolling/ctrl_support/BTActionNode.hpp
```
```cpp
// Copyright (c) 2018 Intel Corporation
//
// Licensed under the Apache License, Version 2.0 (the "License");
// you may not use this file except in compliance with the License.
// You may obtain a copy of the License at
//
//     http://www.apache.org/licenses/LICENSE-2.0
//
// Unless required by applicable law or agreed to in writing, software
// distributed under the License is distributed on an "AS IS" BASIS,
// WITHOUT WARRANTIES OR CONDITIONS OF ANY KIND, either express or implied.
// See the License for the specific language governing permissions and
// limitations under the License.

#ifndef BR2_BT_PATROLLING__CTRL_SUPPORT__BTACTIONNODE_HPP_
#define BR2_BT_PATROLLING__CTRL_SUPPORT__BTACTIONNODE_HPP_

#include <memory>
#include <string>

#include "behaviortree_cpp_v3/action_node.h"
#include "rclcpp/rclcpp.hpp"
#include "rclcpp_action/rclcpp_action.hpp"

namespace br2_bt_patrolling
{

using namespace std::chrono_literals;  // NOLINT

template<class ActionT, class NodeT = rclcpp::Node>
class BtActionNode : public BT::ActionNodeBase
{
public:
  BtActionNode(
    const std::string & xml_tag_name,
    const std::string & action_name,
    const BT::NodeConfiguration & conf)
  : BT::ActionNodeBase(xml_tag_name, conf), action_name_(action_name)
  {
    node_ = config().blackboard->get<typename NodeT::SharedPtr>("node");

    server_timeout_ = 1s;

    // Initialize the input and output messages
    goal_ = typename ActionT::Goal();
    result_ = typename rclcpp_action::ClientGoalHandle<ActionT>::WrappedResult();

    std::string remapped_action_name;
    if (getInput("server_name", remapped_action_name)) {
      action_name_ = remapped_action_name;
    }
    createActionClient(action_name_);

    // Give the derive class a chance to do any initialization
    RCLCPP_INFO(
      node_->get_logger(), "\"%s\" BtActionNode initialized", xml_tag_name.c_str());
  }

  BtActionNode() = delete;
```

br2_bt_patrolling/include/br2_bt_patrolling/ctrl_support/BTActionNode.hpp

```cpp
  virtual ~BtActionNode()
  {
  }

  // Create instance of an action server
  void createActionClient(const std::string & action_name)
  {
    // Now that we have the ROS node to use, create the action client for this BT action
    action_client_ = rclcpp_action::create_client<ActionT>(node_, action_name);

    // Make sure the server is actually there before continuing
    RCLCPP_INFO(
      node_->get_logger(), "Waiting for \"%s\" action server", action_name.c_str());
    action_client_->wait_for_action_server();
  }
  // Any subclass of BtActionNode that accepts parameters must provide a
  // providedPorts method and call providedBasicPorts in it.
  static BT::PortsList providedBasicPorts(BT::PortsList addition)
  {
    BT::PortsList basic = {
      BT::InputPort<std::string>("server_name", "Action server name"),
      BT::InputPort<std::chrono::milliseconds>("server_timeout")
    };
    basic.insert(addition.begin(), addition.end());

    return basic;
  }

  static BT::PortsList providedPorts()
  {
    return providedBasicPorts({});
  }

  // Derived classes can override any of the following methods to hook into the
  // processing for the action: on_tick, on_wait_for_result, and on_success

  // Could do dynamic checks, such as getting updates to values on the blackboard
  virtual void on_tick()
  {
  }

  // There can be many loop iterations per tick. Any opportunity to do something after
  // a timeout waiting for a result that hasn't been received yet
  virtual void on_wait_for_result()
  {
  }

  // Called upon successful completion of the action. A derived class can override this
  // method to put a value on the blackboard, for example.
  virtual BT::NodeStatus on_success()
  {
    return BT::NodeStatus::SUCCESS;
  }

  // Called when a the action is aborted. By default, the node will return FAILURE.
  // The user may override it to return another value, instead.
  virtual BT::NodeStatus on_aborted()
  {
    return BT::NodeStatus::FAILURE;
  }

  // Called when a the action is cancelled. By default, the node will return SUCCESS.
  // The user may override it to return another value, instead.
  virtual BT::NodeStatus on_cancelled()
  {
    return BT::NodeStatus::SUCCESS;
  }

  // The main override required by a BT action
  BT::NodeStatus tick() override
  {
```

```
br2_bt_patrolling/include/br2_bt_patrolling/ctrl_support/BTActionNode.hpp
      // first step to be done only at the beginning of the Action
      if (status() == BT::NodeStatus::IDLE) {
        createActionClient(action_name_);

        // setting the status to RUNNING to notify the BT Loggers (if any)
        setStatus(BT::NodeStatus::RUNNING);

        // user defined callback
        on_tick();

        on_new_goal_received();
      }

      // The following code corresponds to the "RUNNING" loop
      if (rclcpp::ok() && !goal_result_available_) {
        // user defined callback. May modify the value of "goal_updated_"
        on_wait_for_result();

        auto goal_status = goal_handle_->get_status();
        if (goal_updated_ && (goal_status == action_msgs::msg::GoalStatus::STATUS_EXECUTING ||
          goal_status == action_msgs::msg::GoalStatus::STATUS_ACCEPTED))
        {
          goal_updated_ = false;
          on_new_goal_received();
        }

        rclcpp::spin_some(node_->get_node_base_interface());

        // check if, after invoking spin_some(), we finally received the result
        if (!goal_result_available_) {
          // Yield this Action, returning RUNNING
          return BT::NodeStatus::RUNNING;
        }
      }

      switch (result_.code) {
        case rclcpp_action::ResultCode::SUCCEEDED:
          return on_success();

        case rclcpp_action::ResultCode::ABORTED:
          return on_aborted();

        case rclcpp_action::ResultCode::CANCELED:
          return on_cancelled();

        default:
          throw std::logic_error("BtActionNode::Tick: invalid status value");
      }
    }

    // The other (optional) override required by a BT action. In this case, we
    // make sure to cancel the ROS 2 action if it is still running.
    void halt() override
    {
      if (should_cancel_goal()) {
        auto future_cancel = action_client_->async_cancel_goal(goal_handle_);
        if (rclcpp::spin_until_future_complete(
            node_->get_node_base_interface(), future_cancel, server_timeout_) !=
          rclcpp::FutureReturnCode::SUCCESS)
        {
          RCLCPP_ERROR(
            node_->get_logger(),
            "Failed to cancel action server for %s", action_name_.c_str());
        }
      }

      setStatus(BT::NodeStatus::IDLE);
    }

  protected:
```

br2_bt_patrolling/include/br2_bt_patrolling/ctrl_support/BTActionNode.hpp

```cpp
bool should_cancel_goal()
{
  // Shut the node down if it is currently running
  if (status() != BT::NodeStatus::RUNNING) {
    return false;
  }

  rclcpp::spin_some(node_->get_node_base_interface());
  auto status = goal_handle_->get_status();

  // Check if the goal is still executing
  return status == action_msgs::msg::GoalStatus::STATUS_ACCEPTED ||
         status == action_msgs::msg::GoalStatus::STATUS_EXECUTING;
}

void on_new_goal_received()
{
  goal_result_available_ = false;
  auto send_goal_options = typename rclcpp_action::Client<ActionT>::SendGoalOptions();
  send_goal_options.result_callback =
    [this](const typename
      rclcpp_action::ClientGoalHandle<ActionT>::WrappedResult & result) {
      // TODO(#1652): a work around until rcl_action interface is updated
      // if goal ids are not matched, the older goal call this callback so ignore
      //   the result if matched, it must be processed (including aborted)
      if (this->goal_handle_->get_goal_id() == result.goal_id) {
        goal_result_available_ = true;
        result_ = result;
      }
    };

  auto future_goal_handle = action_client_->async_send_goal(goal_, send_goal_options);
  if (rclcpp::spin_until_future_complete(
      node_->get_node_base_interface(), future_goal_handle, server_timeout_) !=
    rclcpp::FutureReturnCode::SUCCESS)
  {
    throw std::runtime_error("send_goal failed");
  }

  goal_handle_ = future_goal_handle.get();
  if (!goal_handle_) {
    throw std::runtime_error("Goal was rejected by the action server");
  }
}

void increment_recovery_count()
{
  int recovery_count = 0;
  config().blackboard->get<int>("number_recoveries", recovery_count);  // NOLINT
  recovery_count += 1;
  config().blackboard->set<int>("number_recoveries", recovery_count);  // NOLINT
}

std::string action_name_;
typename std::shared_ptr<rclcpp_action::Client<ActionT>> action_client_;

// All ROS 2 actions have a goal and a result
typename ActionT::Goal goal_;
bool goal_updated_{false};
bool goal_result_available_{false};
typename rclcpp_action::ClientGoalHandle<ActionT>::SharedPtr goal_handle_;
typename rclcpp_action::ClientGoalHandle<ActionT>::WrappedResult result_;

// The node that will be used for any ROS operations
typename NodeT::SharedPtr node_;

// The timeout value while waiting for response from a server when a
// new action goal is sent or canceled
std::chrono::milliseconds server_timeout_;
};
}  // namespace br2_bt_patrolling
#endif  // BR2_BT_PATROLLING__CTRL_SUPPORT__BTACTIONNODE_HPP_
```

br2_bt_patrolling/include/br2_bt_patrolling/Recharge.hpp

```cpp
#ifndef BR2_BT_PATROLLING__RECHARGE_HPP_
#define BR2_BT_PATROLLING__RECHARGE_HPP_

#include <string>

#include "behaviortree_cpp_v3/behavior_tree.h"
#include "behaviortree_cpp_v3/bt_factory.h"

namespace br2_bt_patrolling
{

class Recharge : public BT::ActionNodeBase
{
public:
  explicit Recharge(
    const std::string & xml_tag_name,
    const BT::NodeConfiguration & conf);

  void halt();
  BT::NodeStatus tick();
  static BT::PortsList providedPorts()
  {
    return BT::PortsList({});
  }

private:
  int counter_;
};

}  // namespace br2_bt_patrolling

#endif  // BR2_BT_PATROLLING__RECHARGE_HPP_
```

br2_bt_patrolling/include/br2_bt_patrolling/GetWaypoint.hpp

```cpp
#ifndef BR2_BT_PATROLLING__GETWAYPOINT_HPP_
#define BR2_BT_PATROLLING__GETWAYPOINT_HPP_

#include <string>
#include <vector>

#include "behaviortree_cpp_v3/behavior_tree.h"
#include "behaviortree_cpp_v3/bt_factory.h"

#include "geometry_msgs/msg/pose_stamped.hpp"

namespace br2_bt_patrolling
{

class GetWaypoint : public BT::ActionNodeBase
{
public:
  explicit GetWaypoint(
    const std::string & xml_tag_name,
    const BT::NodeConfiguration & conf);

  void halt();
  BT::NodeStatus tick();

  static BT::PortsList providedPorts()
  {
    return BT::PortsList(
      {
        BT::InputPort<std::string>("wp_id"),
        BT::OutputPort<geometry_msgs::msg::PoseStamped>("waypoint")
      });
  }

private:
  geometry_msgs::msg::PoseStamped recharge_point_;
```

br2_bt_patrolling/include/br2_bt_patrolling/GetWaypoint.hpp

```cpp
  std::vector<geometry_msgs::msg::PoseStamped> waypoints_;
  static int current_;
};

}  // namespace br2_bt_patrolling

#endif  // BR2_BT_PATROLLING__GETWAYPOINT_HPP_
```

br2_bt_patrolling/include/br2_bt_patrolling/Patrol.hpp

```cpp
#ifndef BR2_BT_PATROLLING__PATROL_HPP_
#define BR2_BT_PATROLLING__PATROL_HPP_

#include <string>

#include "behaviortree_cpp_v3/behavior_tree.h"
#include "behaviortree_cpp_v3/bt_factory.h"

#include "geometry_msgs/msg/twist.hpp"

#include "rclcpp/rclcpp.hpp"

namespace br2_bt_patrolling
{

class Patrol : public BT::ActionNodeBase
{
public:
  explicit Patrol(
    const std::string & xml_tag_name,
    const BT::NodeConfiguration & conf);

  void halt();
  BT::NodeStatus tick();

  static BT::PortsList providedPorts()
  {
    return BT::PortsList({});
  }

private:
  rclcpp::Node::SharedPtr node_;
  rclcpp::Time start_time_;
  rclcpp::Publisher<geometry_msgs::msg::Twist>::SharedPtr vel_pub_;
};

}  // namespace br2_bt_patrolling

#endif  // BR2_BT_PATROLLING__PATROL_HPP_
```

br2_bt_patrolling/include/br2_bt_patrolling/TrackObjects.hpp

```cpp
#ifndef BR2_BT_PATROLLING__TRACKOBJECTS_HPP_
#define BR2_BT_PATROLLING__TRACKOBJECTS_HPP_

#include <string>

#include "geometry_msgs/msg/pose_stamped.hpp"
#include "nav2_msgs/action/navigate_to_pose.hpp"

#include "br2_bt_patrolling/ctrl_support/BTLifecycleCtrlNode.hpp"
#include "behaviortree_cpp_v3/behavior_tree.h"
#include "behaviortree_cpp_v3/bt_factory.h"

namespace br2_bt_patrolling
{
```

```
br2_bt_patrolling/include/br2_bt_patrolling/TrackObjects.hpp
```

```cpp
class TrackObjects : public br2_bt_patrolling::BtLifecycleCtrlNode
{
public:
  explicit TrackObjects(
    const std::string & xml_tag_name,
    const std::string & node_name,
    const BT::NodeConfiguration & conf);
  static BT::PortsList providedPorts()
  {
    return BT::PortsList({});
  }
};

}  // namespace br2_bt_patrolling

#endif  // BR2_BT_PATROLLING__TRACKOBJECTS_HPP_
```

```
br2_bt_patrolling/include/br2_bt_patrolling/Move.hpp
```

```cpp
#ifndef BR2_BT_PATROLLING__MOVE_HPP_
#define BR2_BT_PATROLLING__MOVE_HPP_

#include <string>

#include "geometry_msgs/msg/pose_stamped.hpp"
#include "nav2_msgs/action/navigate_to_pose.hpp"

#include "br2_bt_patrolling/ctrl_support/BTActionNode.hpp"
#include "behaviortree_cpp_v3/behavior_tree.h"
#include "behaviortree_cpp_v3/bt_factory.h"

namespace br2_bt_patrolling
{

class Move : public br2_bt_patrolling::BTActionNode<nav2_msgs::action::NavigateToPose>
{
public:
  explicit Move(
    const std::string & xml_tag_name,
    const std::string & action_name,
    const BT::NodeConfiguration & conf);

  void on_tick() override;
  BT::NodeStatus on_success() override;

  static BT::PortsList providedPorts()
  {
    return {
      BT::InputPort<geometry_msgs::msg::PoseStamped>("goal")
    };
  }
};

}  // namespace br2_bt_patrolling

#endif  // BR2_BT_PATROLLING__MOVE_HPP_
```

```
br2_bt_patrolling/src/br2_bt_patrolling/Patrol.cpp
```

```cpp
#include <string>
#include <iostream>

#include "br2_bt_patrolling/Patrol.hpp"

#include "behaviortree_cpp_v3/behavior_tree.h"

#include "geometry_msgs/msg/twist.hpp"
```

br2_bt_patrolling/src/br2_bt_patrolling/Patrol.cpp

```cpp
#include "rclcpp/rclcpp.hpp"

namespace br2_bt_patrolling
{

using namespace std::chrono_literals;

Patrol::Patrol(
  const std::string & xml_tag_name,
  const BT::NodeConfiguration & conf)
: BT::ActionNodeBase(xml_tag_name, conf)
{
  config().blackboard->get("node", node_);

  vel_pub_ = node_->create_publisher<geometry_msgs::msg::Twist>("/output_vel", 100);
}

void
Patrol::halt()
{
  std::cout << "Patrol halt" << std::endl;
}

BT::NodeStatus
Patrol::tick()
{
  if (status() == BT::NodeStatus::IDLE) {
    start_time_ = node_->now();
  }

  geometry_msgs::msg::Twist vel_msgs;
  vel_msgs.angular.z = 0.5;
  vel_pub_->publish(vel_msgs);

  auto elapsed = node_->now() - start_time_;

  if (elapsed < 15s) {
    return BT::NodeStatus::RUNNING;
  } else {
    return BT::NodeStatus::SUCCESS;
  }
}

}  // namespace br2_bt_patrolling

#include "behaviortree_cpp_v3/bt_factory.h"
BT_REGISTER_NODES(factory)
{
  factory.registerNodeType<br2_bt_patrolling::Patrol>("Patrol");
}
```

br2_bt_patrolling/src/br2_bt_patrolling/Recharge.cpp

```cpp
#include <string>
#include <iostream>
#include <set>

#include "br2_bt_patrolling/Recharge.hpp"

#include "behaviortree_cpp_v3/behavior_tree.h"

namespace br2_bt_patrolling
{

Recharge::Recharge(
  const std::string & xml_tag_name,
  const BT::NodeConfiguration & conf)
: BT::ActionNodeBase(xml_tag_name, conf), counter_(0)
{
}
```

br2_bt_patrolling/src/br2_bt_patrolling/Recharge.cpp

```cpp
void
Recharge::halt()
{
}

BT::NodeStatus
Recharge::tick()
{
  std::cout << "Recharge tick " << counter_ << std::endl;

  if (counter_++ < 50) {
    return BT::NodeStatus::RUNNING;
  } else {
    counter_ = 0;
    config().blackboard->set<float>("battery_level", 100.0f);
    return BT::NodeStatus::SUCCESS;
  }
}

}  // namespace br2_bt_patrolling

#include "behaviortree_cpp_v3/bt_factory.h"
BT_REGISTER_NODES(factory)
{
  factory.registerNodeType<br2_bt_patrolling::Recharge>("Recharge");
}
```

br2_bt_patrolling/src/br2_bt_patrolling/GetWaypoint.cpp

```cpp
#include <string>
#include <iostream>
#include <vector>

#include "br2_bt_patrolling/GetWaypoint.hpp"

#include "behaviortree_cpp_v3/behavior_tree.h"

#include "geometry_msgs/msg/pose_stamped.hpp"

#include "rclcpp/rclcpp.hpp"

namespace br2_bt_patrolling
{

int GetWaypoint::current_ = 0;

GetWaypoint::GetWaypoint(
  const std::string & xml_tag_name,
  const BT::NodeConfiguration & conf)
: BT::ActionNodeBase(xml_tag_name, conf)
{
  rclcpp::Node::SharedPtr node;
  config().blackboard->get("node", node);

  geometry_msgs::msg::PoseStamped wp;
  wp.header.frame_id = "map";
  wp.pose.orientation.w = 1.0;

  // recharge wp
  wp.pose.position.x = 3.67;
  wp.pose.position.y = -0.24;
  recharge_point_ = wp;
```

br2_bt_patrolling/src/br2_bt_patrolling/GetWaypoint.cpp

```cpp
  // wp1
  wp.pose.position.x = 1.07;
```

br2_bt_patrolling/src/br2_bt_patrolling/GetWaypoint.cpp

```cpp
  wp.pose.position.y = -12.38;
  waypoints_.push_back(wp);

  // wp2
  wp.pose.position.x = -5.32;
  wp.pose.position.y = -8.85;
  waypoints_.push_back(wp);

  // wp3
  wp.pose.position.x = -0.56;
  wp.pose.position.y = 0.24;
  waypoints_.push_back(wp);
}

void
GetWaypoint::halt()
{
}

BT::NodeStatus
GetWaypoint::tick()
{
  std::string id;
  getInput("wp_id", id);

  if (id == "recharge") {
    setOutput("waypoint", recharge_point_);
  } else {
    setOutput("waypoint", waypoints_[current_++]);
    current_ = current_ % waypoints_.size();
  }

  return BT::NodeStatus::SUCCESS;
}

}  // namespace br2_bt_patrolling

#include "behaviortree_cpp_v3/bt_factory.h"
BT_REGISTER_NODES(factory)
{
  factory.registerNodeType<br2_bt_patrolling::GetWaypoint>("GetWaypoint");
}
```

br2_bt_patrolling/src/br2_bt_patrolling/Move.cpp

```cpp
#include <string>
#include <iostream>
#include <vector>
#include <memory>

#include "br2_bt_patrolling/Move.hpp"

#include "geometry_msgs/msg/pose_stamped.hpp"
#include "nav2_msgs/action/navigate_to_pose.hpp"

#include "behaviortree_cpp_v3/behavior_tree.h"

namespace br2_bt_patrolling
{

Move::Move(
  const std::string & xml_tag_name,
  const std::string & action_name,
  const BT::NodeConfiguration & conf)
: br2_bt_patrolling::BtActionNode<nav2_msgs::action::NavigateToPose>(xml_tag_name,
    action_name, conf)
{
}
```

```
br2_bt_patrolling/src/br2_bt_patrolling/Move.cpp
```
```cpp
void
Move::on_tick()
{
  geometry_msgs::msg::PoseStamped goal;
  getInput("goal", goal);

  goal_.pose = goal;
}

BT::NodeStatus
Move::on_success()
{
  RCLCPP_INFO(node_->get_logger(), "navigation Suceeded");

  return BT::NodeStatus::SUCCESS;
}

}  // namespace br2_bt_patrolling
#include "behaviortree_cpp_v3/bt_factory.h"
BT_REGISTER_NODES(factory)
{
  BT::NodeBuilder builder =
    [](const std::string & name, const BT::NodeConfiguration & config)
    {
      return std::make_unique<br2_bt_patrolling::Move>(
        name, "navigate_to_pose", config);
    };

  factory.registerBuilder<br2_bt_patrolling::Move>(
    "Move", builder);
}
```

```
br2_bt_patrolling/src/br2_bt_patrolling/BatteryChecker.cpp
```
```cpp
#include <string>
#include <iostream>
#include <algorithm>

#include "br2_bt_patrolling/BatteryChecker.hpp"

#include "behaviortree_cpp_v3/behavior_tree.h"

#include "geometry_msgs/msg/twist.hpp"

#include "rclcpp/rclcpp.hpp"

namespace br2_bt_patrolling
{

using namespace std::chrono_literals;
using namespace std::placeholders;

BatteryChecker::BatteryChecker(
  const std::string & xml_tag_name,
  const BT::NodeConfiguration & conf)
: BT::ConditionNode(xml_tag_name, conf)
{
  config().blackboard->get("node", node_);

  vel_sub_ = node_->create_subscription<geometry_msgs::msg::Twist>(
    "/output_vel", 100, std::bind(&BatteryChecker::vel_callback, this, _1));

  last_reading_time_ = node_->now();
}

void
BatteryChecker::vel_callback(const geometry_msgs::msg::Twist::SharedPtr msg)
{
```

br2_bt_patrolling/src/br2_bt_patrolling/BatteryChecker.cpp

```cpp
  last_twist_ = *msg;
}

void
BatteryChecker::update_battery()
{
  float battery_level;
  if (!config().blackboard->get("battery_level", battery_level)) {
    battery_level = 100.0f;
  }

  float dt = (node_->now() - last_reading_time_).seconds();
  last_reading_time_ = node_->now();

  float vel = sqrt(last_twist_.linear.x * last_twist_.linear.x +
    last_twist_.angular.z * last_twist_.angular.z);
  battery_level = std::max(0.0f, battery_level -(vel * dt * DECAY_LEVEL) - EPSILON * dt);

  config().blackboard->set("battery_level", battery_level);
}
BT::NodeStatus
BatteryChecker::tick()
{
  update_battery();

  float battery_level;
  config().blackboard->get("battery_level", battery_level);

  std::cout << battery_level << std::endl;

  if (battery_level < MIN_LEVEL) {
    return BT::NodeStatus::FAILURE;
  } else {
    return BT::NodeStatus::SUCCESS;
  }
}

}  // namespace br2_bt_patrolling

#include "behaviortree_cpp_v3/bt_factory.h"
BT_REGISTER_NODES(factory)
{
  factory.registerNodeType<br2_bt_patrolling::BatteryChecker>("BatteryChecker");
}
```

br2_bt_patrolling/src/br2_bt_patrolling/TrackObjects.cpp

```cpp
#include <string>
#include <iostream>
#include <vector>
#include <memory>

#include "br2_bt_patrolling/TrackObjects.hpp"

#include "geometry_msgs/msg/pose_stamped.hpp"
#include "nav2_msgs/action/navigate_to_pose.hpp"

#include "behaviortree_cpp_v3/behavior_tree.h"

namespace br2_bt_patrolling
{

TrackObjects::TrackObjects(
  const std::string & xml_tag_name,
  const std::string & action_name,
  const BT::NodeConfiguration & conf)
: br2_bt_patrolling::BtLifecycleCtrlNode(xml_tag_name, action_name, conf)
{
}
```

br2_bt_patrolling/src/br2_bt_patrolling/TrackObjects.cpp

```cpp
}  // namespace br2_bt_patrolling
#include "behaviortree_cpp_v3/bt_factory.h"
BT_REGISTER_NODES(factory)
{
  BT::NodeBuilder builder =
    [](const std::string & name, const BT::NodeConfiguration & config)
    {
      return std::make_unique<br2_bt_patrolling::TrackObjects>(
        name, "/head_tracker", config);
    };

  factory.registerBuilder<br2_bt_patrolling::TrackObjects>(
    "TrackObjects", builder);
}
```

br2_bt_patrolling/src/patrolling main.cpp

```cpp
#include <string>
#include <memory>

#include "behaviortree_cpp_v3/behavior_tree.h"
#include "behaviortree_cpp_v3/bt_factory.h"
#include "behaviortree_cpp_v3/utils/shared_library.h"
#include "behaviortree_cpp_v3/loggers/bt_zmq_publisher.h"

#include "ament_index_cpp/get_package_share_directory.hpp"

#include "rclcpp/rclcpp.hpp"

int main(int argc, char * argv[])
{
  rclcpp::init(argc, argv);

  auto node = rclcpp::Node::make_shared("patrolling_node");

  BT::BehaviorTreeFactory factory;
  BT::SharedLibrary loader;

  factory.registerFromPlugin(loader.getOSName("br2_battery_checker_bt_node"));
  factory.registerFromPlugin(loader.getOSName("br2_patrol_bt_node"));
  factory.registerFromPlugin(loader.getOSName("br2_recharge_bt_node"));
  factory.registerFromPlugin(loader.getOSName("br2_move_bt_node"));
  factory.registerFromPlugin(loader.getOSName("br2_get_waypoint_bt_node"));
  factory.registerFromPlugin(loader.getOSName("br2_track_objects_bt_node"));

  std::string pkgpath = ament_index_cpp::get_package_share_directory("br2_bt_patrolling");
  std::string xml_file = pkgpath + "/behavior_tree_xml/patrolling.xml";

  auto blackboard = BT::Blackboard::create();
  blackboard->set("node", node);
  BT::Tree tree = factory.createTreeFromFile(xml_file, blackboard);

  auto publisher_zmq = std::make_shared<BT::PublisherZMQ>(tree, 10, 2666, 2667);

  rclcpp::Rate rate(10);

  bool finish = false;
  while (!finish && rclcpp::ok()) {
    finish = tree.rootNode()->executeTick() == BT::NodeStatus::SUCCESS;

    rclcpp::spin_some(node);
    rate.sleep();
  }

  rclcpp::shutdown();
  return 0;
}
```

br2_bt_patrolling/tests/bt_action_test.cpp

```cpp
#include <string>
#include <list>
#include <memory>
#include <vector>
#include <set>

#include "behaviortree_cpp_v3/behavior_tree.h"
#include "behaviortree_cpp_v3/bt_factory.h"
#include "behaviortree_cpp_v3/utils/shared_library.h"

#include "ament_index_cpp/get_package_share_directory.hpp"

#include "geometry_msgs/msg/twist.hpp"
#include "nav2_msgs/action/navigate_to_pose.hpp"
#include "lifecycle_msgs/msg/transition.hpp"
#include "lifecycle_msgs/msg/state.hpp"

#include "rclcpp/rclcpp.hpp"
#include "rclcpp_lifecycle/lifecycle_node.hpp"
#include "rclcpp_action/rclcpp_action.hpp"

#include "br2_bt_patrolling/TrackObjects.hpp"

#include "gtest/gtest.h"

using namespace std::placeholders;
using namespace std::chrono_literals;

class VelocitySinkNode : public rclcpp::Node
{
public:
  VelocitySinkNode()
  : Node("VelocitySink")
  {
    vel_sub_ = create_subscription<geometry_msgs::msg::Twist>(
      "/output_vel", 100, std::bind(&VelocitySinkNode::vel_callback, this, _1));
  }

  void vel_callback(geometry_msgs::msg::Twist::SharedPtr msg)
  {
    vel_msgs_.push_back(*msg);
  }

  std::list<geometry_msgs::msg::Twist> vel_msgs_;

private:
  rclcpp::Subscription<geometry_msgs::msg::Twist>::SharedPtr vel_sub_;
};

class Nav2FakeServer : public rclcpp::Node
{
  using NavigateToPose = nav2_msgs::action::NavigateToPose;
  using GoalHandleNavigateToPose = rclcpp_action::ServerGoalHandle<NavigateToPose>;

public:
  Nav2FakeServer()
  : Node("nav2_fake_server_node") {}

  void start_server()
  {
    move_action_server_ = rclcpp_action::create_server<NavigateToPose>(
      shared_from_this(),
      "navigate_to_pose",
      std::bind(&Nav2FakeServer::handle_goal, this, _1, _2),
      std::bind(&Nav2FakeServer::handle_cancel, this, _1),
      std::bind(&Nav2FakeServer::handle_accepted, this, _1));
  }

private:
  rclcpp_action::Server<NavigateToPose>::SharedPtr move_action_server_;
```

```cpp
// br2_bt_patrolling/tests/bt_action_test.cpp
  rclcpp_action::GoalResponse handle_goal(
    const rclcpp_action::GoalUUID & uuid,
    std::shared_ptr<const NavigateToPose::Goal> goal)
  {
    return rclcpp_action::GoalResponse::ACCEPT_AND_EXECUTE;
  }
  rclcpp_action::CancelResponse handle_cancel(
    const std::shared_ptr<GoalHandleNavigateToPose> goal_handle)
  {
    return rclcpp_action::CancelResponse::ACCEPT;
  }

  void handle_accepted(const std::shared_ptr<GoalHandleNavigateToPose> goal_handle)
  {
    std::thread{std::bind(&Nav2FakeServer::execute, this, _1), goal_handle}.detach();
  }

  void execute(const std::shared_ptr<GoalHandleNavigateToPose> goal_handle)
  {
    auto feedback = std::make_shared<NavigateToPose::Feedback>();
    auto result = std::make_shared<NavigateToPose::Result>();

    auto start = now();

    while ((now() - start) < 5s) {
      feedback->distance_remaining = 5.0 - (now() - start).seconds();
      goal_handle->publish_feedback(feedback);
    }

    goal_handle->succeed(result);
  }
};
class StoreWP : public BT::ActionNodeBase
{
public:
  explicit StoreWP(
    const std::string & xml_tag_name,
    const BT::NodeConfiguration & conf)
  : BT::ActionNodeBase(xml_tag_name, conf) {}

  void halt() {}
  BT::NodeStatus tick()
  {
    waypoints_.push_back(getInput<geometry_msgs::msg::PoseStamped>("in").value());
    return BT::NodeStatus::SUCCESS;
  }

  static BT::PortsList providedPorts()
  {
    return BT::PortsList(
    {
      BT::InputPort<geometry_msgs::msg::PoseStamped>("in")
    });
  }

  static std::vector<geometry_msgs::msg::PoseStamped> waypoints_;
};

std::vector<geometry_msgs::msg::PoseStamped> StoreWP::waypoints_;

TEST(bt_action, recharge_btn)
{
  auto node = rclcpp::Node::make_shared("recharge_btn_node");

  BT::BehaviorTreeFactory factory;
  BT::SharedLibrary loader;

  factory.registerFromPlugin(loader.getOSName("br2_recharge_bt_node"));

  std::string xml_bt =
```

br2_bt_patrolling/tests/bt_action_test.cpp

```cpp
    R"(
    <root main_tree_to_execute = "MainTree" >
      <BehaviorTree ID="MainTree">
          <Recharge    name="recharge"/>
      </BehaviorTree>
    </root>)";

  auto blackboard = BT::Blackboard::create();
  blackboard->set("node", node);
  BT::Tree tree = factory.createTreeFromText(xml_bt, blackboard);

  rclcpp::Rate rate(10);

  bool finish = false;
  while (!finish && rclcpp::ok()) {
    finish = tree.rootNode()->executeTick() == BT::NodeStatus::SUCCESS;
    rate.sleep();
  }

  float battery_level;
  ASSERT_TRUE(blackboard->get("battery_level", battery_level));
  ASSERT_NEAR(battery_level, 100.0f, 0.0000001);
}

TEST(bt_action, patrol_btn)
{
  auto node = rclcpp::Node::make_shared("patrol_btn_node");
  auto node_sink = std::make_shared<VelocitySinkNode>();

  BT::BehaviorTreeFactory factory;
  BT::SharedLibrary loader;

  factory.registerFromPlugin(loader.getOSName("br2_patrol_bt_node"));

  std::string xml_bt =
    R"(
    <root main_tree_to_execute = "MainTree" >
      <BehaviorTree ID="MainTree">
          <Patrol     name="patrol"/>
      </BehaviorTree>
    </root>)";

  auto blackboard = BT::Blackboard::create();
  blackboard->set("node", node);
  BT::Tree tree = factory.createTreeFromText(xml_bt, blackboard);

  rclcpp::Rate rate(10);

  bool finish = false;
  int counter = 0;
  while (!finish && rclcpp::ok()) {
    finish = tree.rootNode()->executeTick() == BT::NodeStatus::SUCCESS;
    rclcpp::spin_some(node_sink->get_node_base_interface());
    rate.sleep();
  }

  ASSERT_FALSE(node_sink->vel_msgs_.empty());
  ASSERT_NEAR(node_sink->vel_msgs_.size(), 150, 2);

  geometry_msgs::msg::Twist & one_twist = node_sink->vel_msgs_.front();

  ASSERT_GT(one_twist.angular.z, 0.1);
  ASSERT_NEAR(one_twist.linear.x, 0.0, 0.0000001);
}

TEST(bt_action, move_btn)
{
  auto node = rclcpp::Node::make_shared("move_btn_node");
  auto nav2_fake_node = std::make_shared<Nav2FakeServer>();

  nav2_fake_node->start_server();
```

```cpp
// br2_bt_patrolling/tests/bt_action_test.cpp

  bool finish = false;
  std::thread t([&]() {
      while (!finish) {rclcpp::spin_some(nav2_fake_node);}
    });

  BT::BehaviorTreeFactory factory;
  BT::SharedLibrary loader;

  factory.registerFromPlugin(loader.getOSName("br2_move_bt_node"));

  std::string xml_bt =
    R"(
    <root main_tree_to_execute = "MainTree" >
      <BehaviorTree ID="MainTree">
        <Move    name="move" goal="{goal}"/>
      </BehaviorTree>
    </root>)";

  auto blackboard = BT::Blackboard::create();
  blackboard->set("node", node);

  geometry_msgs::msg::PoseStamped goal;
  blackboard->set("goal", goal);

  BT::Tree tree = factory.createTreeFromText(xml_bt, blackboard);

  rclcpp::Rate rate(10);

  int counter = 0;
  while (!finish && rclcpp::ok()) {
    finish = tree.rootNode()->executeTick() == BT::NodeStatus::SUCCESS;
    rate.sleep();
  }

  t.join();
}

TEST(bt_action, get_waypoint_btn)
{
  auto node = rclcpp::Node::make_shared("get_waypoint_btn_node");

  rclcpp::spin_some(node);

  {
    BT::BehaviorTreeFactory factory;
    BT::SharedLibrary loader;

    factory.registerFromPlugin(loader.getOSName("br2_get_waypoint_bt_node"));

    std::string xml_bt =
      R"(
      <root main_tree_to_execute = "MainTree" >
        <BehaviorTree ID="MainTree">
          <GetWaypoint    name="recharge" wp_id="{id}" waypoint="{waypoint}"/>
        </BehaviorTree>
      </root>)";

    auto blackboard = BT::Blackboard::create();
    blackboard->set("node", node);
    blackboard->set<std::string>("id", "recharge");

    BT::Tree tree = factory.createTreeFromText(xml_bt, blackboard);

    rclcpp::Rate rate(10);

    bool finish = false;
    int counter = 0;
    while (!finish && rclcpp::ok()) {
      finish = tree.rootNode()->executeTick() == BT::NodeStatus::SUCCESS;
      counter++;
```

br2_bt_patrolling/tests/bt_action_test.cpp

```cpp
    rate.sleep();
  }

  auto point = blackboard->get<geometry_msgs::msg::PoseStamped>("waypoint");

  ASSERT_EQ(counter, 1);
  ASSERT_NEAR(point.pose.position.x, 3.67, 0.0000001);
  ASSERT_NEAR(point.pose.position.y, -0.24, 0.0000001);
}

{
  BT::BehaviorTreeFactory factory;
  BT::SharedLibrary loader;

  factory.registerNodeType<StoreWP>("StoreWP");
  factory.registerFromPlugin(loader.getOSName("br2_get_waypoint_bt_node"));

  std::string xml_bt =
    R"(
    <root main_tree_to_execute = "MainTree" >
      <BehaviorTree ID="MainTree">
        <Sequence name="root_sequence">
          <GetWaypoint    name="wp1" wp_id="next" waypoint="{waypoint}"/>
          <StoreWP in="{waypoint}"/>
          <GetWaypoint    name="wp2" wp_id="next" waypoint="{waypoint}"/>
          <StoreWP in="{waypoint}"/>
          <GetWaypoint    name="wp3" wp_id="" waypoint="{waypoint}"/>
          <StoreWP in="{waypoint}"/>
          <GetWaypoint    name="wp4" wp_id="recharge" waypoint="{waypoint}"/>
          <StoreWP in="{waypoint}"/>
          <GetWaypoint    name="wp5" wp_id="wp1" waypoint="{waypoint}"/>
          <StoreWP in="{waypoint}"/>
          <GetWaypoint    name="wp6" wp_id="wp2" waypoint="{waypoint}"/>
          <StoreWP in="{waypoint}"/>
          <GetWaypoint    name="wpt" waypoint="{waypoint}"/>
          <StoreWP in="{waypoint}"/>
        </Sequence>
      </BehaviorTree>
    </root>)";

  auto blackboard = BT::Blackboard::create();
  blackboard->set("node", node);

  BT::Tree tree = factory.createTreeFromText(xml_bt, blackboard);

  rclcpp::Rate rate(10);

  bool finish = false;
  while (!finish && rclcpp::ok()) {
    finish = tree.rootNode()->executeTick() == BT::NodeStatus::SUCCESS;
    rate.sleep();
  }

  const auto & waypoints = StoreWP::waypoints_;
  ASSERT_EQ(waypoints.size(), 7);
  ASSERT_NEAR(waypoints[0].pose.position.x, 1.07, 0.0000001);
  ASSERT_NEAR(waypoints[0].pose.position.y, -12.38, 0.0000001);
  ASSERT_NEAR(waypoints[1].pose.position.x, -5.32, 0.0000001);
  ASSERT_NEAR(waypoints[1].pose.position.y, -8.85, 0.0000001);
  ASSERT_NEAR(waypoints[2].pose.position.x, -0.56, 0.0000001);
  ASSERT_NEAR(waypoints[2].pose.position.y, 0.24, 0.0000001);

  ASSERT_NEAR(waypoints[3].pose.position.x, 3.67, 0.0000001);
  ASSERT_NEAR(waypoints[3].pose.position.y, -0.24, 0.0000001);

  ASSERT_NEAR(waypoints[4].pose.position.x, 1.07, 0.0000001);
  ASSERT_NEAR(waypoints[4].pose.position.y, -12.38, 0.0000001);
  ASSERT_NEAR(waypoints[5].pose.position.x, -5.32, 0.0000001);
  ASSERT_NEAR(waypoints[5].pose.position.y, -8.85, 0.0000001);
  ASSERT_NEAR(waypoints[6].pose.position.x, -0.56, 0.0000001);
  ASSERT_NEAR(waypoints[6].pose.position.y, 0.24, 0.0000001);
}
```

```cpp
// br2_bt_patrolling/tests/bt_action_test.cpp
}

TEST(bt_action, battery_checker_btn)
{
  auto node = rclcpp::Node::make_shared("battery_checker_btn_node");
  auto vel_pub = node->create_publisher<geometry_msgs::msg::Twist>("/output_vel", 100);

  BT::BehaviorTreeFactory factory;
  BT::SharedLibrary loader;

  factory.registerFromPlugin(loader.getOSName("br2_battery_checker_bt_node"));
  factory.registerFromPlugin(loader.getOSName("br2_patrol_bt_node"));

  std::string xml_bt =
    R"(
    <root main_tree_to_execute = "MainTree" >
      <BehaviorTree ID="MainTree">
        <ReactiveSequence>
            <BatteryChecker    name="battery_checker"/>
            <Patrol            name="patrol"/>
        </ReactiveSequence>
      </BehaviorTree>
    </root>)";

  auto blackboard = BT::Blackboard::create();
  blackboard->set("node", node);
  BT::Tree tree = factory.createTreeFromText(xml_bt, blackboard);

  rclcpp::Rate rate(10);
  geometry_msgs::msg::Twist vel;
  vel.linear.x = 0.8;

  bool finish = false;
  int counter = 0;
  while (!finish && rclcpp::ok()) {
    finish = tree.rootNode()->executeTick() == BT::NodeStatus::SUCCESS;

    vel_pub->publish(vel);

    rclcpp::spin_some(node);
    rate.sleep();
  }

  float battery_level;
  ASSERT_TRUE(blackboard->get("battery_level", battery_level));
  ASSERT_NEAR(battery_level, 94.6, 1.0);
}

TEST(bt_action, track_objects_btn_1)
{
  auto node = rclcpp::Node::make_shared("track_objects_btn_node");
  auto node_head_tracker = rclcpp_lifecycle::LifecycleNode::make_shared("head_tracker");

  bool finish = false;
  std::thread t([&]() {
      while (!finish) {rclcpp::spin_some(node_head_tracker->get_node_base_interface());}
    });

  BT::NodeConfiguration conf;
  conf.blackboard = BT::Blackboard::create();
  conf.blackboard->set("node", node);
  br2_bt_patrolling::BtLifecycleCtrlNode bt_node("TrackObjects", "head_tracker", conf);

  bt_node.change_state_client_ = bt_node.createServiceClient<lifecycle_msgs::srv::
    ChangeState>("/head_tracker/change_state");
  ASSERT_TRUE(bt_node.change_state_client_->service_is_ready());

  bt_node.get_state_client_ = bt_node.createServiceClient<lifecycle_msgs::srv::GetState>(
    "/head_tracker/get_state");
  ASSERT_TRUE(bt_node.get_state_client_->service_is_ready());
  auto start = node->now();
```

br2_bt_patrolling/tests/bt_action_test.cpp

```cpp
  rclcpp::Rate rate(10);
  while (rclcpp::ok() && (node->now() - start) < 1s) {
    rclcpp::spin_some(node);
    rate.sleep();
  }

  ASSERT_EQ(bt_node.get_state(), lifecycle_msgs::msg::State::PRIMARY_STATE_UNCONFIGURED);
  bt_node.ctrl_node_state_ = lifecycle_msgs::msg::State::PRIMARY_STATE_UNCONFIGURED;
  ASSERT_FALSE(bt_node.set_state(lifecycle_msgs::msg::State::PRIMARY_STATE_ACTIVE));

  node_head_tracker->trigger_transition(lifecycle_msgs::msg::Transition::
    TRANSITION_CONFIGURE);

  start = node->now();
  while (rclcpp::ok() && (node->now() - start) < 1s) {
    rclcpp::spin_some(node);
    rate.sleep();
  }

  bt_node.ctrl_node_state_ = bt_node.get_state();

  ASSERT_TRUE(bt_node.set_state(lifecycle_msgs::msg::State::PRIMARY_STATE_ACTIVE));
  ASSERT_EQ(bt_node.get_state(), lifecycle_msgs::msg::State::PRIMARY_STATE_ACTIVE);

  start = node->now();
  while (rclcpp::ok() && (node->now() - start) < 1s) {
    rclcpp::spin_some(node);
    rate.sleep();
  }

  bt_node.ctrl_node_state_ = bt_node.get_state();

  ASSERT_TRUE(bt_node.set_state(lifecycle_msgs::msg::State::PRIMARY_STATE_INACTIVE));
  ASSERT_EQ(bt_node.get_state(), lifecycle_msgs::msg::State::PRIMARY_STATE_INACTIVE);

  finish = true;
  t.join();
}

TEST(bt_action, track_objects_btn_2)
{
  auto node = rclcpp::Node::make_shared("track_objects_btn_node");
  auto node_head_tracker = rclcpp_lifecycle::LifecycleNode::make_shared("head_tracker");

  bool finish = false;
  std::thread t([&]() {
      while (!finish) {rclcpp::spin_some(node_head_tracker->get_node_base_interface());}
    });

  BT::NodeConfiguration conf;
  conf.blackboard = BT::Blackboard::create();
  conf.blackboard->set("node", node);
  br2_bt_patrolling::BtLifecycleCtrlNode bt_node("TrackObjects", "head_tracker", conf);

  node_head_tracker->trigger_transition(lifecycle_msgs::msg::Transition::
    TRANSITION_CONFIGURE);

  rclcpp::Rate rate(10);
  auto start = node->now();
  while (rclcpp::ok() && (node->now() - start) < 1s) {
    rclcpp::spin_some(node);
    rate.sleep();
  }

  ASSERT_EQ(bt_node.tick(), BT::NodeStatus::RUNNING);

  ASSERT_TRUE(bt_node.change_state_client_->service_is_ready());
  ASSERT_TRUE(bt_node.get_state_client_->service_is_ready());

  ASSERT_EQ(bt_node.get_state(), lifecycle_msgs::msg::State::PRIMARY_STATE_ACTIVE);
```

```
br2_bt_patrolling/tests/bt_action_test.cpp
```
```cpp
  ASSERT_EQ(bt_node.tick(), BT::NodeStatus::RUNNING);

  bt_node.halt();

  start = node->now();
  while (rclcpp::ok() && (node->now() - start) < 1s) {
    rclcpp::spin_some(node);
    rate.sleep();
  }

  ASSERT_EQ(bt_node.get_state(), lifecycle_msgs::msg::State::PRIMARY_STATE_INACTIVE);

  finish = true;
  t.join();
}

TEST(bt_action, track_objects_btn_3)
{
  auto node = rclcpp::Node::make_shared("track_objects_btn_node");
  auto node_head_tracker = rclcpp_lifecycle::LifecycleNode::make_shared("head_tracker");

  node_head_tracker->trigger_transition(lifecycle_msgs::msg::Transition::
    TRANSITION_CONFIGURE);

  bool finish = false;
  std::thread t([&]() {
    while (!finish) {rclcpp::spin_some(node_head_tracker->get_node_base_interface());}
  });

  BT::BehaviorTreeFactory factory;
  BT::SharedLibrary loader;

  factory.registerFromPlugin(loader.getOSName("br2_track_objects_bt_node"));

  std::string xml_bt =
    R"(
    <root main_tree_to_execute = "MainTree" >
      <BehaviorTree ID="MainTree">
        <KeepRunningUntilFailure>
            <TrackObjects    name="track_objects"/>
        </KeepRunningUntilFailure>
      </BehaviorTree>
    </root>)";

  auto blackboard = BT::Blackboard::create();
  blackboard->set("node", node);
  auto start = node->now();
  rclcpp::Rate rate(10);

  {
    BT::Tree tree = factory.createTreeFromText(xml_bt, blackboard);

    ASSERT_EQ(
      node_head_tracker->get_current_state().id(),
      lifecycle_msgs::msg::State::PRIMARY_STATE_INACTIVE);

    while (rclcpp::ok() && (node->now() - start) < 1s) {
      tree.rootNode()->executeTick() == BT::NodeStatus::RUNNING;

      rclcpp::spin_some(node);
      rate.sleep();
    }
    ASSERT_EQ(
      node_head_tracker->get_current_state().id(),
      lifecycle_msgs::msg::State::PRIMARY_STATE_ACTIVE);
  }

  start = node->now();
  while (rclcpp::ok() && (node->now() - start) < 1s) {
    rclcpp::spin_some(node);
```

br2_bt_patrolling/tests/bt_action_test.cpp

```cpp
    rate.sleep();
  }

  ASSERT_EQ(
    node_head_tracker->get_current_state().id(),
    lifecycle_msgs::msg::State::PRIMARY_STATE_INACTIVE);

  finish = true;
  t.join();
}

TEST(bt_action, move_track_btn)
{
  auto node = rclcpp::Node::make_shared("move_btn_node");
  auto nav2_fake_node = std::make_shared<Nav2FakeServer>();
  auto node_head_tracker = rclcpp_lifecycle::LifecycleNode::make_shared("head_tracker");

  node_head_tracker->trigger_transition(lifecycle_msgs::msg::Transition::
    TRANSITION_CONFIGURE);

  nav2_fake_node->start_server();

  rclcpp::executors::SingleThreadedExecutor exe;
  exe.add_node(nav2_fake_node);
  exe.add_node(node_head_tracker->get_node_base_interface());
  bool finish = false;
  std::thread t([&]() {
      while (!finish) {exe.spin_some();}
    });

  BT::BehaviorTreeFactory factory;
  BT::SharedLibrary loader;

  factory.registerFromPlugin(loader.getOSName("br2_move_bt_node"));
  factory.registerFromPlugin(loader.getOSName("br2_track_objects_bt_node"));

  std::string xml_bt =
    R"(
    <root main_tree_to_execute = "MainTree" >
      <BehaviorTree ID="MainTree">
        <Parallel success_threshold="1" failure_threshold="1">
          <TrackObjects    name="track_objects"/>
          <Move    name="move" goal="{goal}"/>
        </Parallel>
      </BehaviorTree>
    </root>)";

  auto blackboard = BT::Blackboard::create();
  blackboard->set("node", node);

  geometry_msgs::msg::PoseStamped goal;
  blackboard->set("goal", goal);

  BT::Tree tree = factory.createTreeFromText(xml_bt, blackboard);
  ASSERT_EQ(
    node_head_tracker->get_current_state().id(),
    lifecycle_msgs::msg::State::PRIMARY_STATE_INACTIVE);

  rclcpp::Rate rate(10);
  auto start = node->now();
  auto finish_tree = false;
  while (rclcpp::ok() && (node->now() - start) < 1s) {
    finish_tree = tree.rootNode()->executeTick() == BT::NodeStatus::SUCCESS;

    rclcpp::spin_some(node);
    rate.sleep();
  }

  ASSERT_FALSE(finish_tree);
  ASSERT_EQ(
    node_head_tracker->get_current_state().id(),
```

br2_bt_patrolling/tests/bt_action_test.cpp

```cpp
    lifecycle_msgs::msg::State::PRIMARY_STATE_ACTIVE);

  while (rclcpp::ok() && !finish_tree) {
    finish_tree = tree.rootNode()->executeTick() == BT::NodeStatus::SUCCESS;

    rclcpp::spin_some(node);
    rate.sleep();
  }

  start = node->now();
  while (rclcpp::ok() && (node->now() - start) < 1s) {
    rclcpp::spin_some(node);
    rate.sleep();
  }

  ASSERT_EQ(
    node_head_tracker->get_current_state().id(),
    lifecycle_msgs::msg::State::PRIMARY_STATE_INACTIVE);

  finish = true;
  t.join();
}

int main(int argc, char ** argv)
{
  rclcpp::init(argc, argv);

  testing::InitGoogleTest(&argc, argv);
  return RUN_ALL_TESTS();
}
```

br2_bt_patrolling/tests/CMakeLists.txt

```
ament_add_gtest(bt_action_test bt_action_test.cpp)
ament_target_dependencies(bt_action_test ${dependencies})
target_link_libraries(bt_action_test br2_track_objects_bt_node)
```

REFERENCES
参考文献

[1] Rodney A. Brooks. Elephants don't play chess. *Robotics and Autonomous Systems*, 6(1):3–15, 1990. Designing Autonomous Agents.

[2] Brian Gerkey, Richard T Vaughan, Andrew Howard, et al. The player/stage project: Tools for multi-robot and distributed sensor systems. In *Proceedings of the 11th international conference on advanced robotics*, volume 1, pages 317–323. Citeseer, 2003.

[3] Steven Macenski, Francisco Martin, Ruffin White, and Jonatan Gint'es Clavero. The marathon 2: A navigation system. In *2020 IEEE/RSJ International Conference on Intelligent Robots and Systems (IROS)*, 2020.

[4] Alejandro Marzinotto, Michele Colledanchise, Christian Smith, and Petter Ögren. Towards a unified behavior trees framework for robot control. In *2014 IEEE International Conference on Robotics and Automation (ICRA)*, pages 5420–5427, 2014.

[5] Giorgio Metta, Paul Fitzpatrick, and Lorenzo Natale. Yarp: Yet another robot platform. *International Journal of Advanced Robotic Systems*, 3(1):8, 2006.

[6] Michael Montemerlo, Nicholas Roy, and Sebastian Thrun. Perspectives on standardization in mobile robot programming: The carnegie mellon navigation (carmen) toolkit. In *Proceedings 2003 IEEE/RSJ International Conference on Intelligent Robots and Systems (IROS 2003)(Cat. No. 03CH37453)*, volume 3, pages 2436–2441. IEEE, 2003.

[7] Morgan Quigley, Brian Gerkey, Ken Conley, Josh Faust, Tully Foote, Jeremy Leibs, Eric Berger, Rob Wheeler, and Andrew Ng. Ros: an open-source robot operating system. In *Proc. of the IEEE Intl. Conf. on Robotics and Automation (ICRA) Workshop on Open Source Robotics*, Kobe, Japan, May 2009.

[8] Dirk Thomas, William Woodall, and Esteve Fernandez. Next-generation ros: Building on dds. In OSRF, editor, *ROSCon 2014*, 2014.

[9] Sebastian Thrun, Dieter Fox, Wolfram Burgard, and Frank Dellaert. Robust monte carlo localization for mobile robots. *Artificial Intelligence*, 128(1):99–141, 2001.

图 2.11

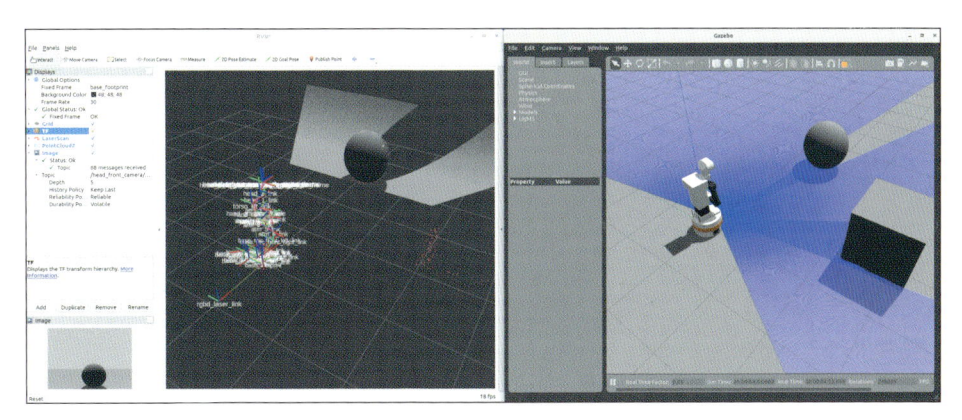

图 2.13

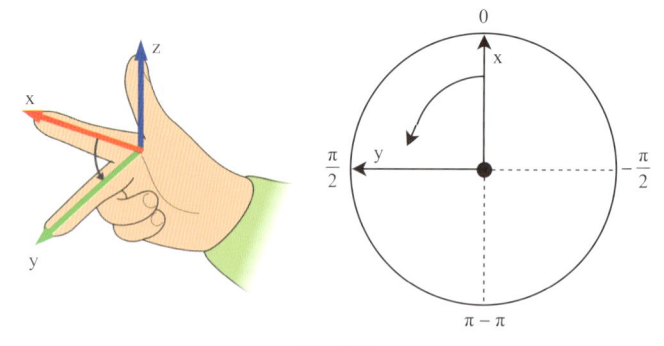

图 3.2

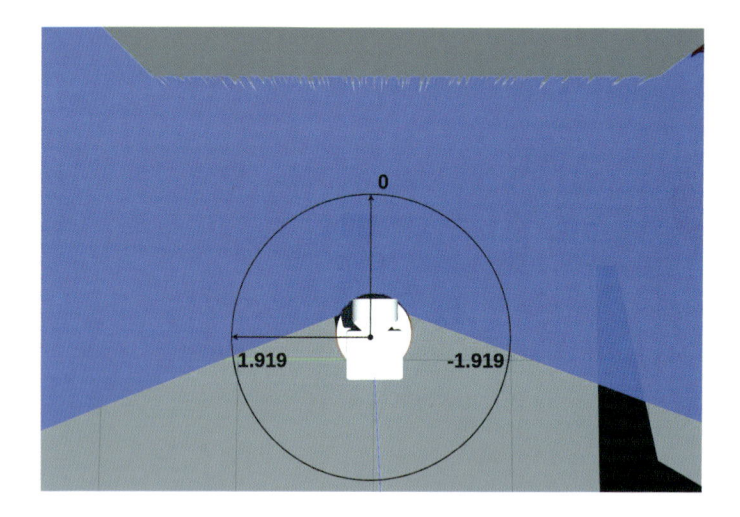

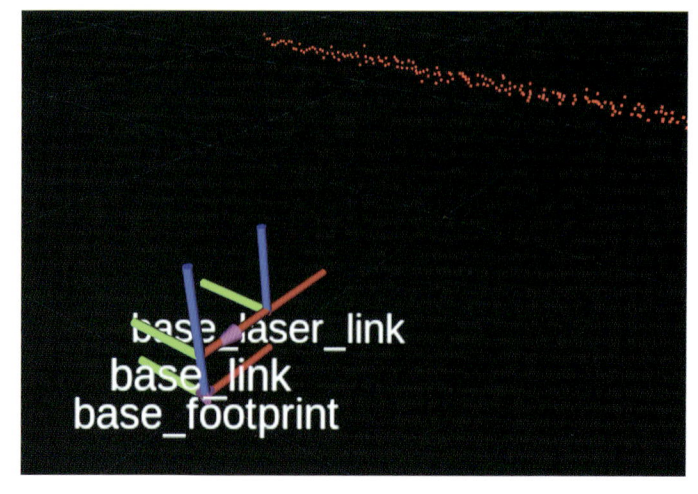

图 3.3

图 4.1

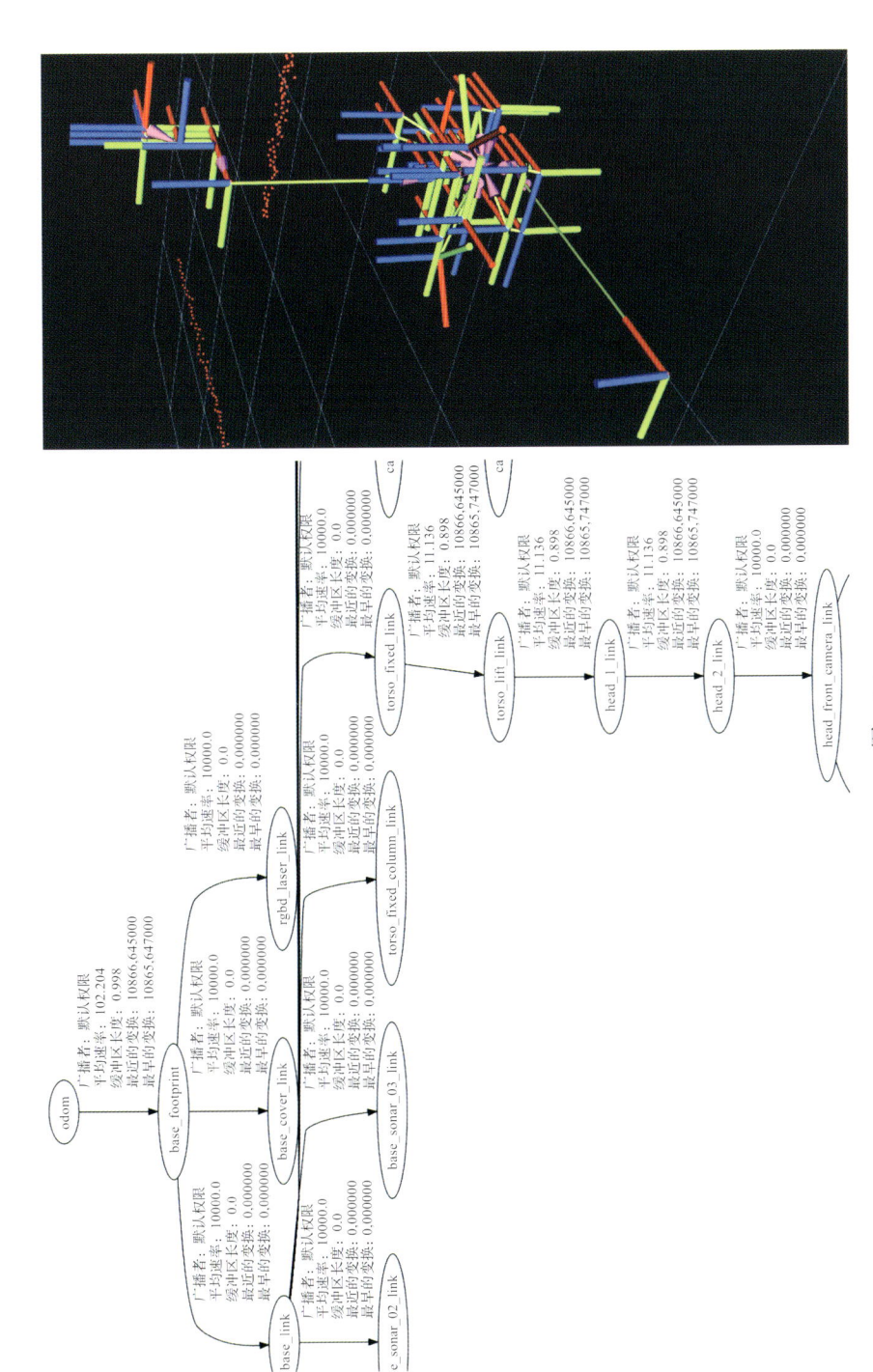

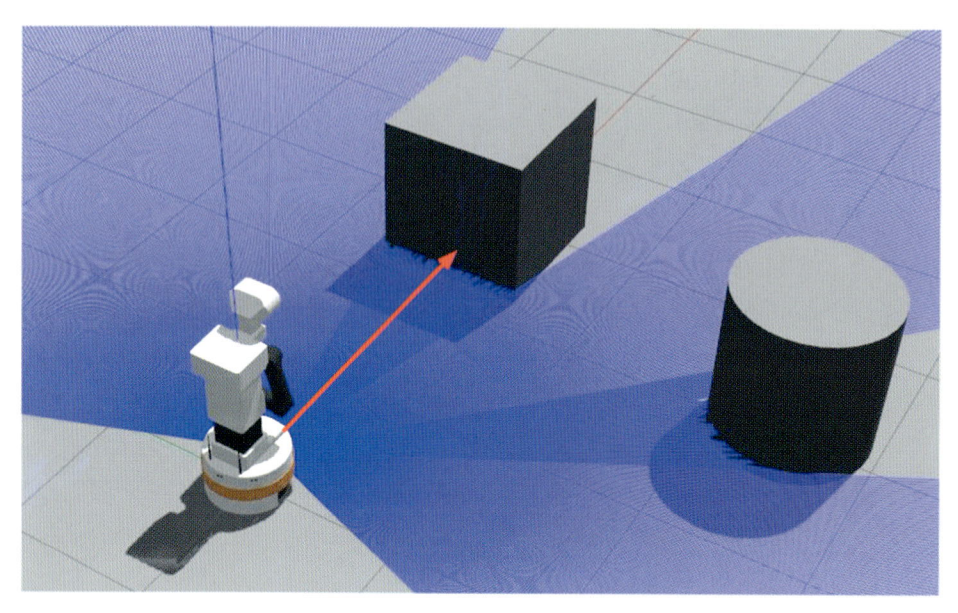

图 4.3

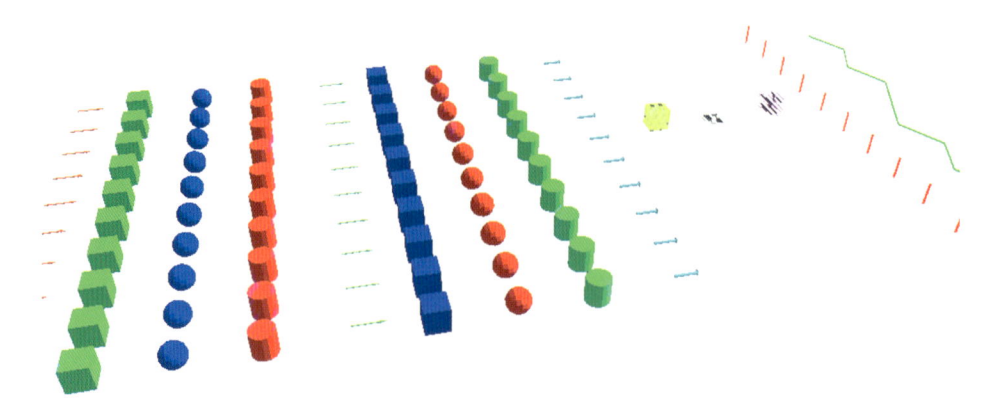

图 4.4

图 4.6

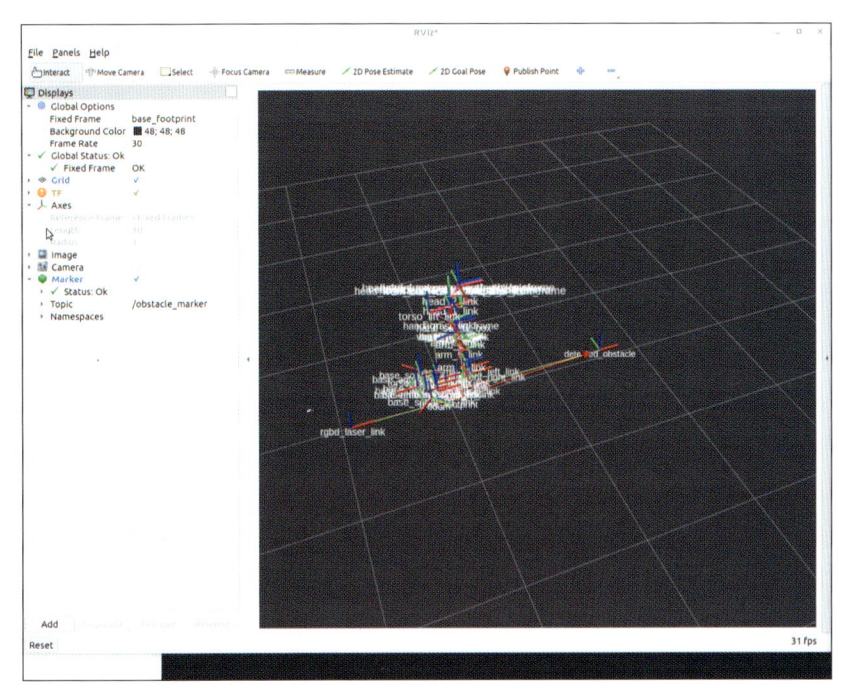

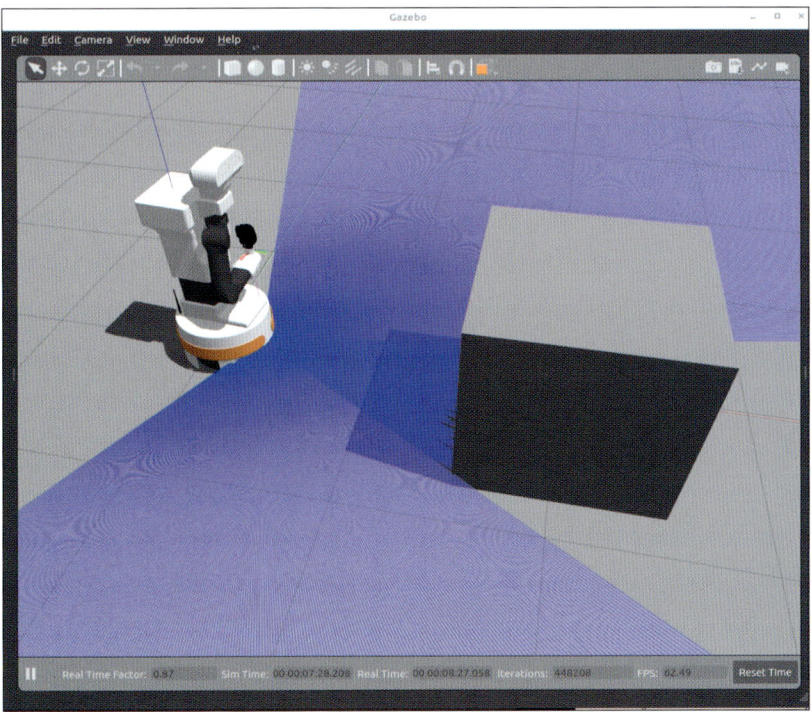

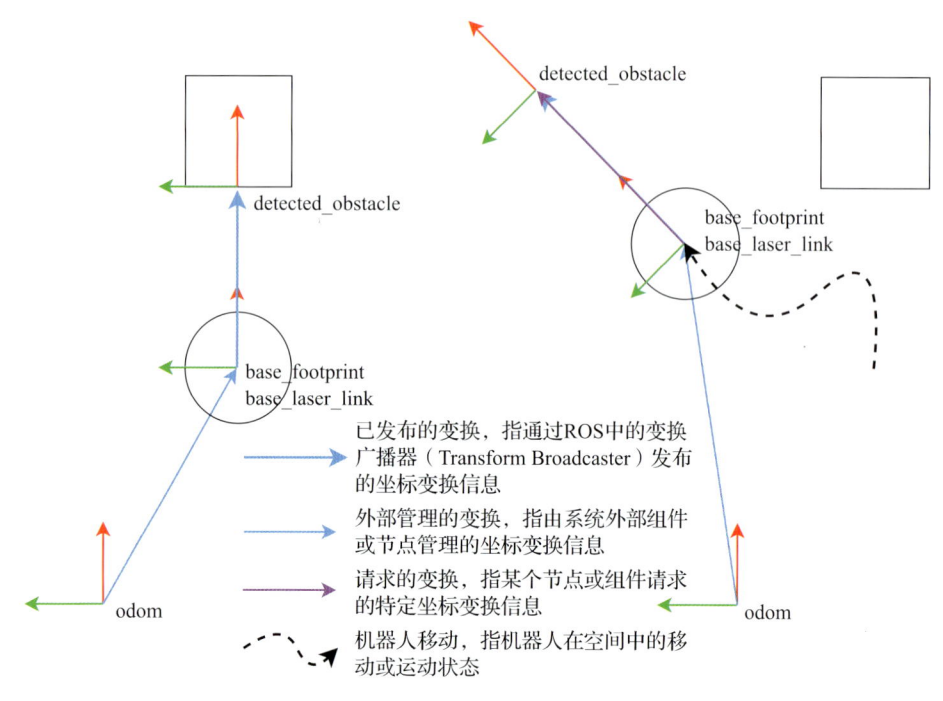

图　4.7

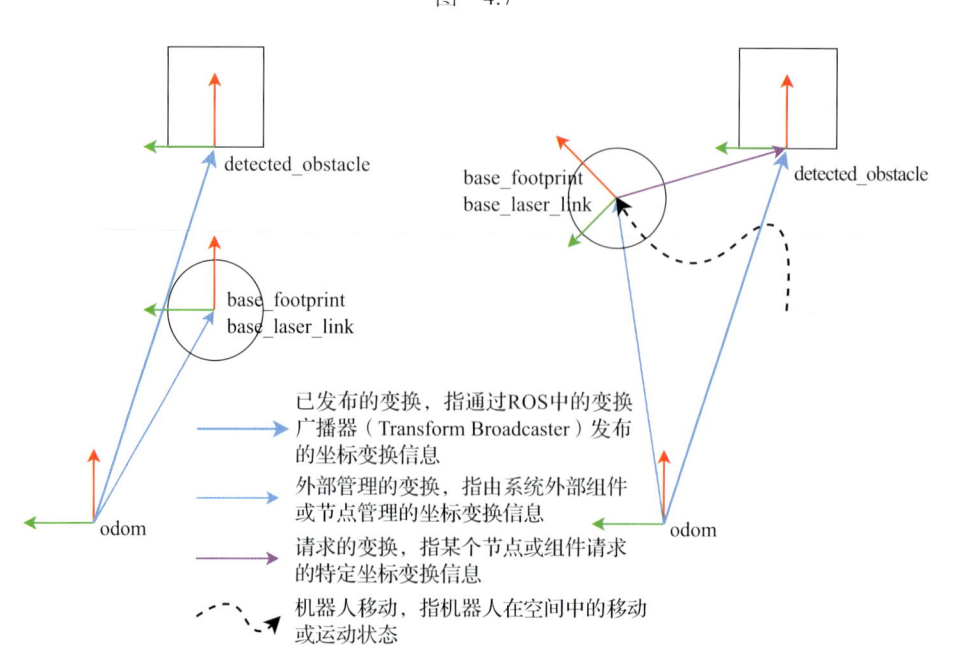

图　4.8

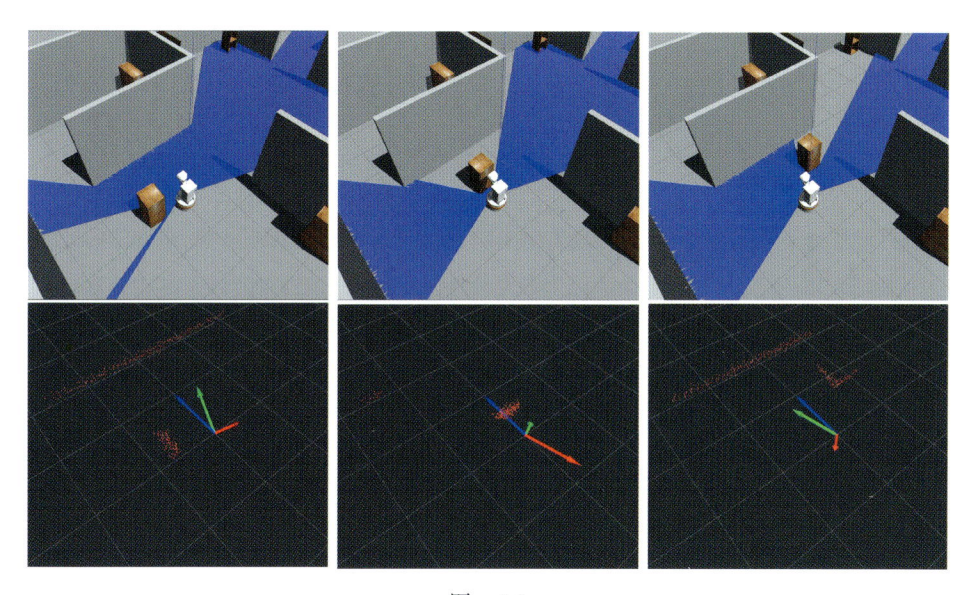

图 5.1

图 5.3

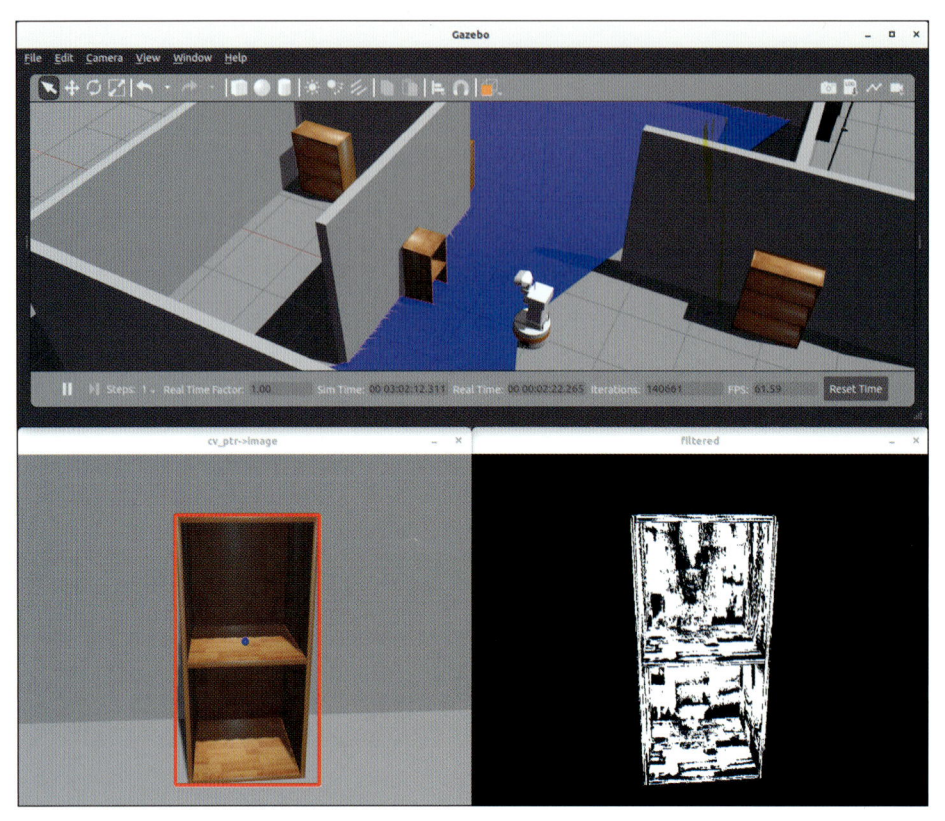

图 5.4

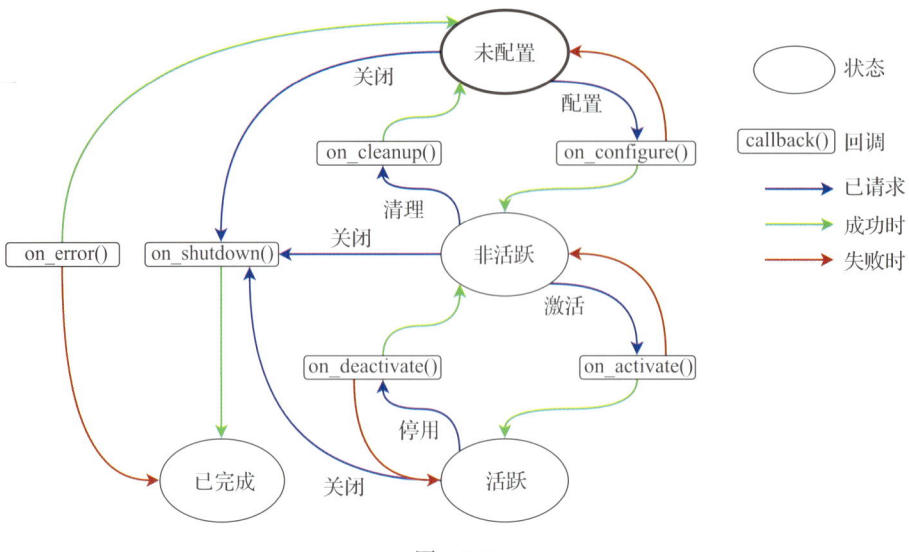

图 5.9

图 5.11

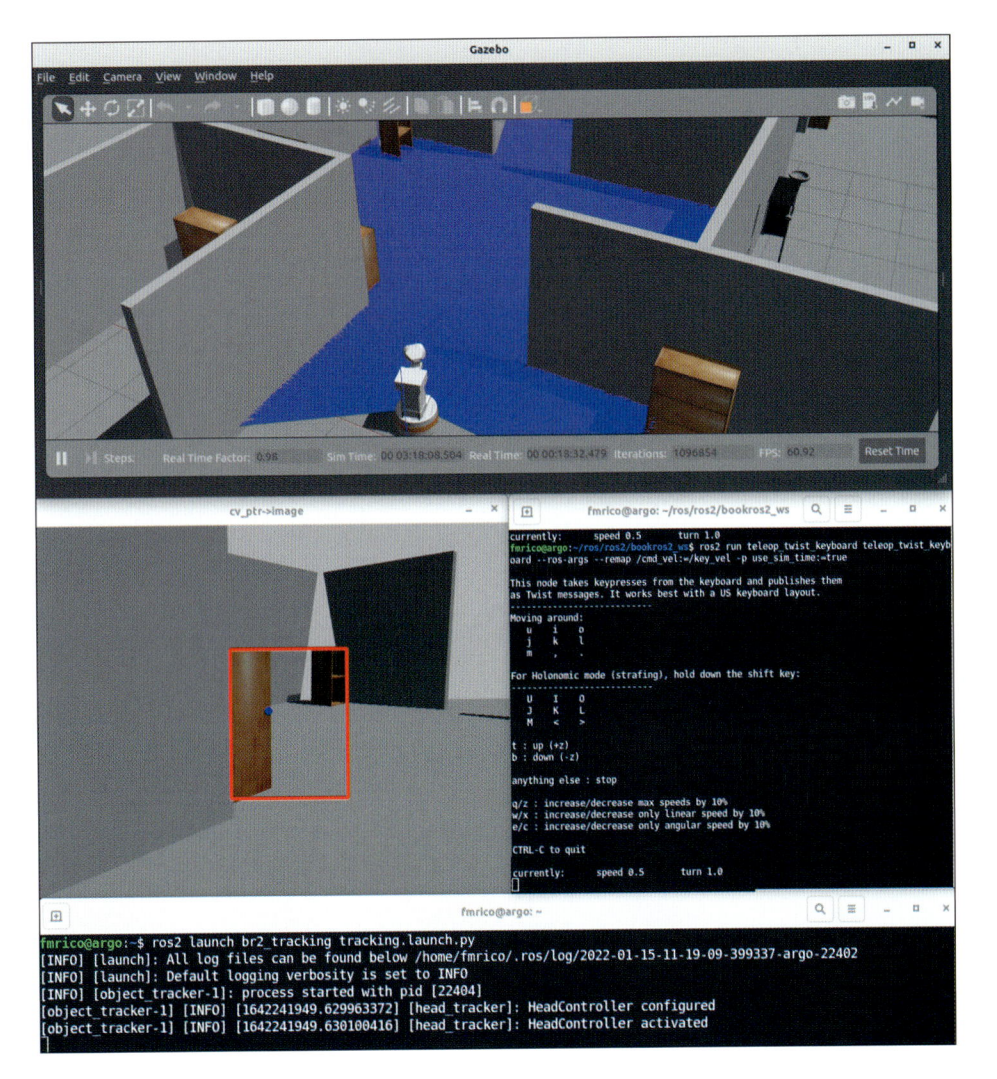

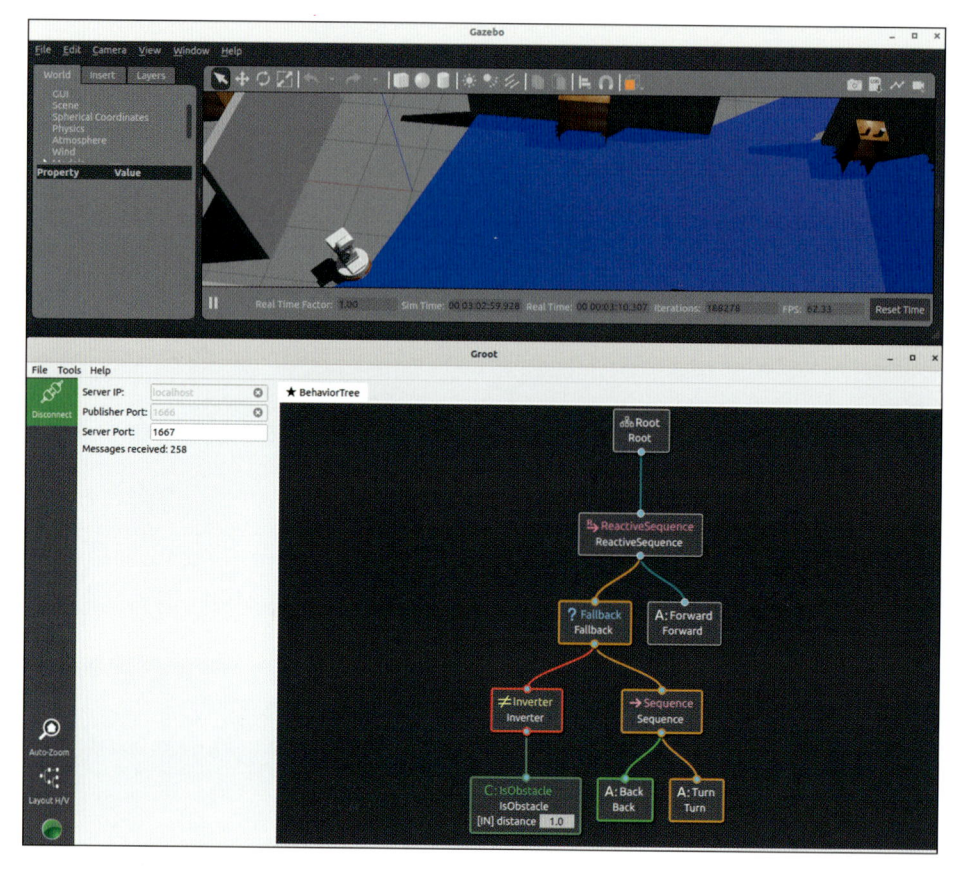

图 6.13

图 6.18

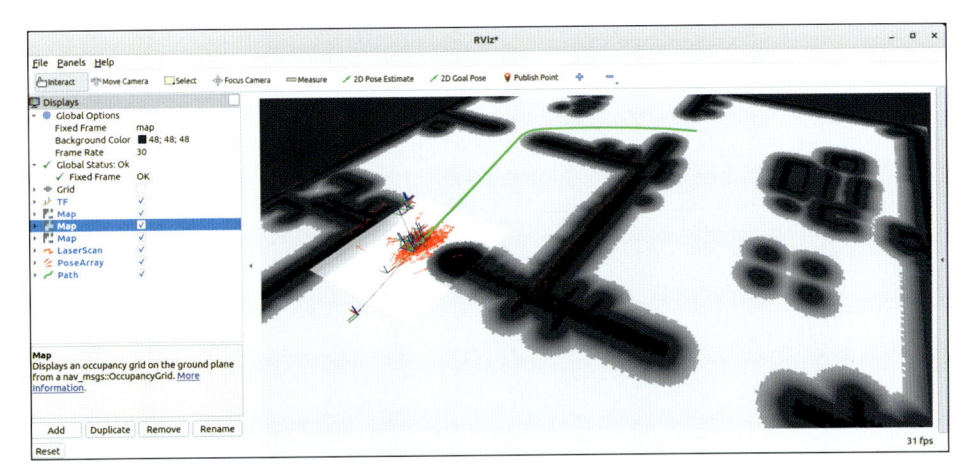

图 6.18 （续）

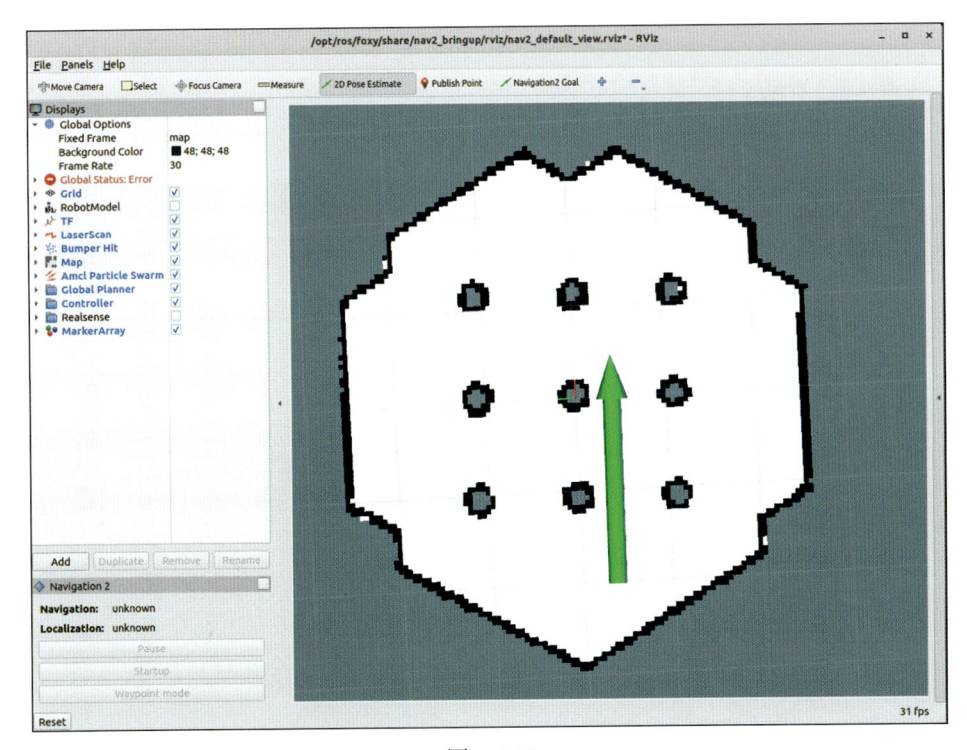

图 6.19

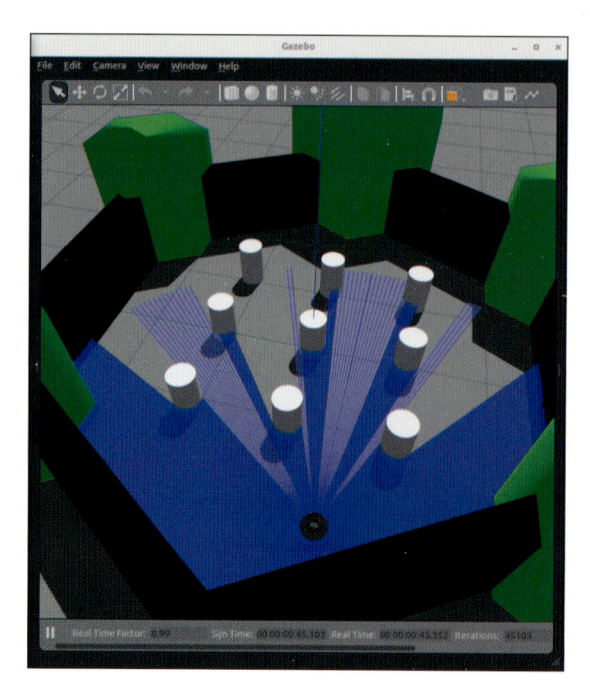

图 6.19 （续）

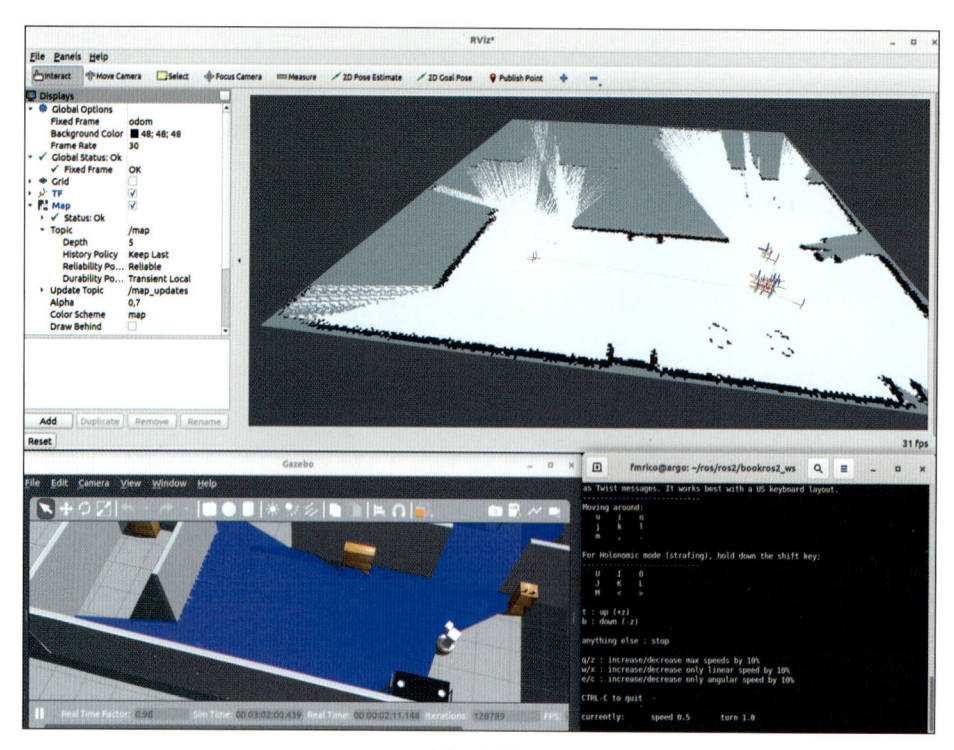

图 6.20